The Unemployment Crisis

The Unemployment Crisis

RICHARD LAYARD   STEPHEN NICKELL
RICHARD JACKMAN

OXFORD UNIVERSITY PRESS
1994

Oxford University Press, Walton Street, Oxford OX2 6DP

Oxford New York Toronto
Delhi Bombay Calcutta Madras Karachi
Kuala Lumpur Singapore Hong Kong Tokyo
Nairobi Dar es Salaam Cape Town
Melbourne Auckland Madrid

and associated companies in
Berlin Ibadan

Oxford is a trade mark of Oxford University Press

Published in the United States
by Oxford University Press Inc., New York

© R. Layard, S. Nickell, R. Jackman 1994

British Library Cataloguing in Publication Data
Data available
ISBN 0–19–877395–1
ISBN 0–19–877394–3 (Pbk)

Library of Congress Cataloging in Publication Data
Data available

1 3 5 7 9 10 8 6 4 2

Set by Hope Services (Abingdon) Ltd.
Printed in Great Britain
on acid-free paper by
Biddles Ltd
Guildford & King's Lynn

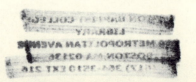

To the millions who suffer
through want of work

CONTENTS

PREFACE

Once again in the early 1990s world unemployment is high and rising—as it was in the early 1980s. The issue is particularly acute in Europe, where the European Union has identified it as the overriding challenge to be met. But in the US too unemployment rates are very high for many groups in the community, leading President Clinton to sponsor an international Jobs Summit.

The immediate crisis is related to the world recession of the early 1990s. But much more serious than the business cycle downturn is the average level of unemployment over the whole business cycle. Over the last ten years unemployment in the European Union has averaged nearly 10 per cent of the workforce, which is a much more serious matter than the fluctuations of unemployment around that average.

Our book is mainly addressed to the issue of why unemployment is so high over a run of years and secondly to the sources of fluctuations. We began this work in the early 1980s, in the belief that the unemployment issue was unlikely to go away. When unemployment fell in the late 1980s, many people thought the problem was being solved, and one leading economist now shaping US policy told us that we had better publish a book soon before the problem had altogether disappeared.

We finally published our first book in 1991 by which time unemployment was again rising towards record levels in many countries. The book was called *Unemployment: Macroeconomic Performance and the Labour Market*, and it appears to have been well received. It was nominated as an outstanding academic book of the year by Choice, in the USA.

The book was rather long (618 pages) however and rather expensive for students. We were therefore very pleased when the publisher proposed that we should publish the Overview section as a separate book, suitably updated.

The present book gives in a succinct form our main arguments and findings. It is written for the general reader who is concerned with the issues, and knows only the most elementary economics. (The argument can be followed even if the maths is skipped.)

For what ultimately matters in the world is what is in the minds of politicians, administrators, and voters. We would not have written the book unless we hoped that it would affect how they think.

Our general approach

Some explanation is needed of our general approach and how it advances understanding of the unemployment problem. Clearly no economy can function well without *some* unemployment. But do we really need so much unemployment? And, if not, how can it be reduced?

To answer these questions, we must first understand how unemployment comes about and why it changes. We develop a general framework of analysis, and then use it to explain the history of our times. Unemployment depends on so many different factors that it is not easy to find a single coherent framework for analysing how they interact. Yet without such a framework, it is difficult to refute the apparent plausibility of a thousand quack remedies.

An adequate framework requires a new combination of macroeconomics with a detailed micro analysis of the labour market. Traditionally, macroeconomics has concerned itself with how temporary shocks make unemployment fluctuate in the short term around its average level, while labour economics has focused on what determines that average level—factors such as unemployment insurance, labour mobility, and the like. But it has become more and more obvious that the average level itself varies greatly between decades, with previous unemployment exerting a persistent effect on subsequent unemployment. To explain this persistence requires new micro foundations of macroeconomics, going far beyond the influences considered in the 1970s.

A key issue is the role of the employed 'insiders' and the unemployed 'outsiders' in the labour market. How do they affect wage pressure and thus set limits to non-inflationary growth? The employed insiders want to have wages set in their own interest, with little regard to the interests of the un-employed outsiders. But the outsiders still have a role. If they search less hard for work or are unsuited to the jobs available, this reduces the effective supply of labour and thus increases wage pressure. Long-term unemployment diminishes the effectiveness of the individual as a supplier of labour and hence it becomes quite easy for a relatively small shock, like an oil price rise, to have long-lasting effects.

But there are many other issues. To fit them all in, we develop a single integrated view of the labour market, which explains both the stock of unemployed people and the flows into and out of unemployment, as well as the evolution of wage and price inflation. It allows for union bargaining, efficiency wages, unemployment insurance, labour mobility, and many other influences. It draws on microeconomic and macroeconomic evidence, and provides a convincing explana-tion of the astonishing movements of unemployment and inflation that have occurred in the post-war world.

The analysis we present is in many ways original. This is bound to be the case, since recent events have so often been inconsistent with old explanations, and since many basic issues still lack an adequate analytical framework. But we have also aimed to incorporate the best of existing knowledge. In this sense we have tried to write a book that is simultaneously a contribution to new social thought and a textbook.

Our work would have been impossible without the generous help of our many friends and colleagues at LSE and else-where. They are listed in our previous book. Throughout this work, the Centre has been financed by the Economic and Social Research Council, the Department of Employment, and the Esmee Fairbairn Charitable Trust. Their support has been invaluable. But the chief support has been from our families. Thank you all.

London and Oxford.   December 1993.
R. Layard
S. Nickell
R. Jackman

1

Facts to be Explained

UNEMPLOYMENT matters. It generally reduces output and aggregate income. It increases inequality, since the unemployed lose more than the employed. It erodes human capital. And, finally, it involves psychic costs. People need to be needed. Though unemployment increases leisure, the value of this is largely offset by the pain of rejection.

So we have to explain why unemployment occurs, how it changes over time, and why it affects some kinds of people and not others. We can then suggest policies that will make things better.

Let us begin with some of the key facts that need to be explained.

1. Unemployment fluctuates over time

Some of these fluctuations are short-term changes which get reversed quite quickly. But there are also big secular changes (see Fig. 1). The 1960s were a period of very low unemployment. Since then unemployment has risen in most countries. The rise has been much worse in the European Community (EC) than anywhere else, with unemployment increasing in every year between 1973 and 1986 (from 3 to 11 per cent). After 1986 European unemployment fell very slowly until the early 1990s when it began to rise again.

Unemployment rate (%)

Unemployment rate (%)

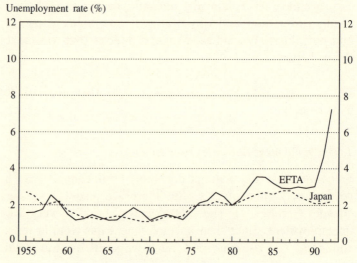

Fig. 1. *Unemployment, 1955–1992.*
EFTA (the European Free Trade Area) includes Norway, Sweden, Finland, Austria, and Switzerland. Detailed annual data for each country are in Annex 6.

Sources: see Annex 6.

Unemployment rate (%)

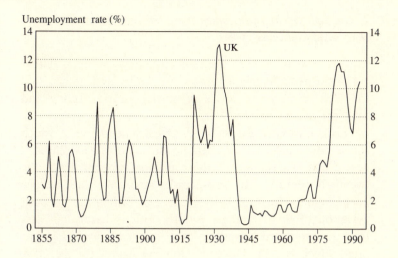

Unemployment rate (%)

Fig. 2. *Unemployment since the nineteenth century.*

Sources: *UK*: Feinstein (1972), chained to data in Annex 6. OECD series. *USA*: 1890–1954: US Census, *Historical Statistics of the United States* (1976). 1900 Series D85–86, chained to 1955–1992 series in Annex 6.

2. Unemployment varies much more between business cycles than within business cycles

This is true of almost all countries. For example, unemployment rose hugely between the 1920s and 1930s, and then fell to very low levels in most countries during and after the Second World War.

This is illustrated for the USA and Britain in Fig. 2, which shows how much average unemployment varies between business cycles. To summarize this variation, we can divide the twentieth century into half-decades and take the average unemployment for each half-decade. For Britain the standard deviation of these averages is 3.16. This is hardly any less than the standard deviation of the *annual* unemployment rates, which is 3.36. The corresponding figures for the USA are 4.29 and 4.88. Thus, most of the annual variation 'comes from' the long-frequency fluctuations between half-decades rather than from the short-frequency fluctuations within half-decades.[1] Conventional business cycles account for relatively little of the history of unemployment.

The reasons for this are a central issue of this book. In our view they stem from two sources: first, there are long-period changes in social institutions; and, second, big shocks to the system (such as oil price rises or major wars) have long-lasting effects.

The main social institutions that affect unemployment are the unemployment benefit system and the system of wage determination. In Europe unemployment benefit systems generally became more generous financially and more readily available up to around 1980. This did not happen in the USA. In addition, the position of the unions became increasingly strong in Europe up to around 1980. Union membership grew in many countries, while it was falling in the USA. On most indices, militancy grew. For example, from 1968 onwards (the year of the Paris riots) the number of industrial conflicts rose sharply (see Fig. 3). Even before the oil shocks, increased militancy was making it difficult to contain inflation without rising unemployment.

Conflicts per 1,000 non-agricultural workers

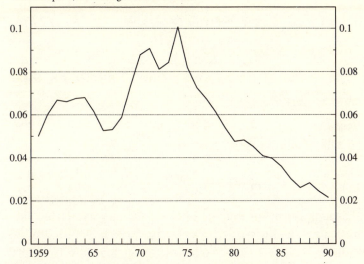

Fig. 3. *Industrial conflicts in the OECD, 1959–1990.*
Sources: ILO, *Yearbook of Labour Statistics*; OECD, *Labour Force Statistics*.

Index (1980=100)

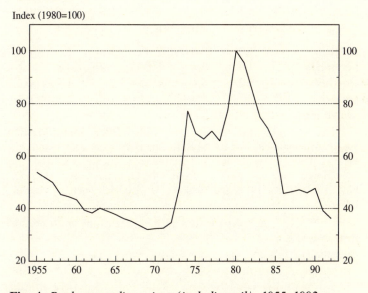

Fig. 4. *Real commodity prices (including oil), 1955–1992.*
Sources: UN, *Statistical Papers* Series M, no. 82, and *Monthly Bulletin*; IMF, *International Financial Statistics*; OECD, *Main Economic Indicators*.

Chapter 1

Table 1 *Percentage of labour force unemployed, 1979 and 1990*

	1990			1979		
	All	Under 1 year	Over 1 year	All	Under 1 year	Over 1 year
Belgium	7.2	1.6	5.6	8.2	3.4	4.8
Denmark	9.6	6.8	2.8	6.2	—	—
France	8.9	5.4	3.5	5.9	4.1	1.8
Germany	4.9	2.5	2.4	3.2	2.6	0.6
Ireland	13.4	4.6	8.8	7.1	4.8	2.3
Italy	7.0	2.1	4.9	5.2	3.3	1.9
Netherlands	7.5	3.7	3.8	5.4	3.9	1.5
Portugal	5.1	2.5	2.6	4.8	—	—
Spain	15.9	6.6	9.3	8.5	6.1	2.4
UK	6.8	3.8	3.0	5.0	3.8	1.3
Australia	6.9	5.3	1.6	6.2	5.1	1.1
New Zealand	7.8	—	—	1.9	—	—
Canada	8.1	7.6	0.5	7.4	7.1	0.3
USA	5.4	5.1	0.3	5.8	5.6	0.2
Japan	2.1	1.7	0.4	2.1	1.7	0.4
Austria	3.2	2.8	0.4	1.8	1.6	0.2
Finland	3.4	2.8	0.6	5.9	4.8	1.1
Norway	5.2	4.6	0.6	2.0	1.9	0.1
Sweden	1.5	1.4	0.1	1.7	1.6	0.1
Switzerland	1.8	—	—	0.9	—	—

Source: Unemployment rates have so far as possible been standardized, as described in Annex 6. Percentage of unemployed who are unemployed over one year is from OECD, *Employment Outlook*, 1985, Table H (for 1979) and 1990, Table M (which refers to 1988 or 1989).

Notes: Detailed country series for unemployment and inflation are given in Annex 6.

Throughout this book, 'Germany' refers to 'West Germany'.

However, it was the big commodity price shocks of 1973–4 and 1979–80, shown in Fig. 4, that gave the sharpest impulse to inflation. And the ensuing efforts of governments to disinflate then led to the further rises in unemployment. Europe, as a major importer of raw materials, suffered much more from the commodity price rises than did the USA, which is much more self-sufficient. But what surprised everybody was the extraordinary persistence of European unemployment in the 1980s, and the fact that inflation fell so slowly despite mass unemployment. In our view, a key to understanding this is the emergence of long-term unemployment.

3. The rise in European unemployment has been associated with a massive increase in long-term unemployment (see Table 1)

In most European countries the proportion of workers entering unemployment is quite small: it is much lower than in the USA and has risen little. The huge difference is in the duration of unemployment: nearly half of Europe's unemployed have now been out of work for over a year. As we shall show, long-term unemployment reduces the effectiveness of the unemployed as potential fillers of vacancies. Once long-term unemployment has taken root, it has a very weak tendency to correct itself. This helps to explain our next fact.

4. In many countries the level of unemployment has risen sharply relative to the level of vacancies

This suggests either an increase in mismatch (which we question) or a failure of the unemployed to seek work as effectively as before.

5. Despite all this, unemployment is untrended over the very long term (see Fig. 2)

This is a key point. It suggests that ultimately there are

very powerful mechanisms at work which have forced the
number of jobs to respond to huge changes that have
occurred in the numbers of people wanting work. It also
suggests that in the long term productivity and taxes have
no impact on unemployment.

These are the main time-series facts about unemployment. We turn now to cross-sectional differences.

6. Unemployment differs greatly between countries (see Table 1)

Among industrial countries it is worst in the countries
of the EC, while the other Western European (EFTA)
countries (Norway, Sweden, Finland, Austria, and Switzerland) and Japan have been remarkably unaffected (see Fig.
1) (except very recently in the case of EFTA). This appears
to be due to differences in social institutions, with the latter
countries having highly corporatist wage-setting arrangements and/or shorter entitlements to benefits (combined in
Sweden with major training and employment programmes
for the unemployed). These arrangements both inhibited
unemployment's original rise and ensured that unemployment did not persist. In the USA there was by contrast a
big rise in unemployment in the early 1980s, but, with
unemployment benefits running out after six months, this
could not persist.

The differences in unemployment rates in Table 1 are
quite genuine. People are defined as unemployed if they are
not working but are available for work and have taken
specific steps to find work within the last month. This is the
standard OECD definition, and the data are generally got
by household surveys such as the EC Labour Force Survey
or the US Current Population Survey. Unemployed people
do of course differ in the intensity with which they seek
work and in the type of work they are willing to accept.
We shall discuss this issue at length. But it in no way inval-
idates the concept of unemployment, any more than the

concept of tallness is invalidated by the fact that, if we defined tall as 'over 6 feet', some people are even taller.

7. *Few unemployed people have deliberately chosen to become unemployed*

In the USA about a half have lost their last job, a quarter have re-entered the labour force after an interval, and over 10 per cent are looking for their first job. Figures for the UK are similar. Only a small minority become unemployed by quitting their last job. Thus, the issue of whether unemployment is in any sense voluntary arises mainly in relation to the duration of unemployment rather than the inflow into it.

8. *Unemployment differs greatly between age-groups, occupations, regions, and races*

As Table 2 shows, young people are much more likely to be unemployed than older people. In some countries like Italy and Spain the differences are truly astounding. And it is clear that countries differ less in the 'core' unemployment of adult males than in youth unemployment or female unemployment.

The most important difference in unemployment rates is between occupations. The rate for semi- and unskilled workers is four to five times higher than that for professional and managerial workers. Over three-quarters of unemployed men are manual workers. Thus, the theory of unemployment has to focus on the labour market for manual workers. The labour market experiences of economists will not throw much light on the subject.

The challenge is to find a consistent and plausible framework which explains the facts. Needless to say, the most plausible framework is one in which the actions of firms and individuals are described in terms that they would themselves recognize.

Table 2 *Percentage of labour force unemployed, 1987*

	All workers (1)	Over 25		Under 25	
		Men (2)	Women (3)	Men (4)	Women (5)
Belgium	11.0	5.6	15.3	16.0	27.1
Denmark	7.8	5.2	9.4	9.3	11.9
France	10.5	6.4	10.1	19.6	27.9
Germany	6.2	5.1	7.5	6.1	8.5
Greece	7.4	3.8	6.7	15.5	35.1
Ireland	17.5	13.5	18.5	27.2	22.6
Italy	7.9	2.3	6.5	21.0	30.1
Netherlands	9.6	6.8	11.7	14.2	14.3
Portugal	7.0	3.3	5.6	13.1	21.5
Spain	20.1	11.9	16.8	39.9	50.1
UK	10.2	8.8	8.0	16.9	14.6
Australia	8.0	5.6	6.1	15.0	14.5
New Zealand	4.1	1.9	2.4	6.1	5.5
Canada	8.8	7.0	8.4	14.9	12.5
USA	6.1	4.8	4.8	12.6	11.7
Japan	2.8	2.6	2.4	5.4	5.0
Austria	3.8	3.4	3.7	4.4	4.7
Finland	5.0	5.0	3.8	9.7	8.1
Norway	2.1	1.8	1.5	3.8	3.9
Sweden	1.9	1.4	1.5	4.4	4.0
Switzerland	1.8	—	—	—	—

Sources: For total unemployment see Annex 6. Age analysis is: *EC*, Eurostat, Series 3C, *Employment and Unemployment*, 1988, Table IV/1, all figures being multiplied by ratio of col. (1) to the Eurostat total; *Others*: ILO, *Yearbook of Labour Statistics*, 1988, Tables 1 and 9B, multiplied by ratio of col. (1) to ILO total.

2

Our Broad Approach

In developing a framework, we start from the fact that, when buoyant demand reduces unemployment (at least relative to recently experienced levels), inflationary pressure develops. Firms start bidding against each other for labour, and workers feel more confident in pressing wage claims. If the inflationary pressure is too great, inflation starts spiralling upwards: higher wage rises lead to higher price rises, leading to still higher wage rises, and so on. This is the wage–price spiral.

The outcome is illustrated for the OECD as a whole in Fig. 5. Panel (*a*) shows the level of economic activity measured by the (detrended) proportion of the labour force in work—in other words, the (detrended) employment rate. It shows clearly the pattern of boom and slump over the last quarter-century. Panel (*b*) shows the inflation rate (GDP deflator). In each boom inflation rose and in each slump it fell. Panel (*c*) therefore shows the relation between the change in the inflation rate and the level of employment. The association between the two is clear.[2]

But increasing inflation can be sustained only by continued monetary and/or fiscal injections. If financial policy is stable, with nominal income growing at a constant rate, rising inflation will in due course lead to rising unemployment. Eventually the higher unemployment will stop inflation rising, and both unemployment and inflation will stabilize.

The level of unemployment at which inflation stabilizes is the *equilibrium* level of unemployment. This concept of equilibrium has nothing to do with the concept of 'market-

Employment rate (minus trend) %

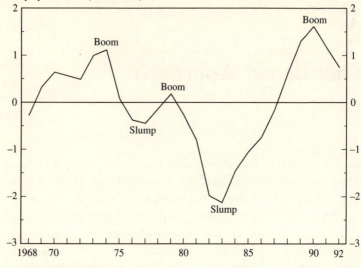

(a) Employment rate

Inflation rate (annual) %

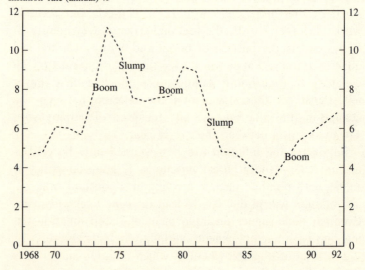

(b) Inflation

Employment rate (minus trend)   Change in inflation

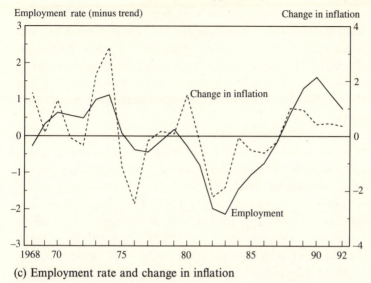

(c) Employment rate and change in inflation

Fig. 5. *Unemployment–inflation trade-off in the OECD.*
Inflation is change in GDP deflator.
Source: OECD.

clearing', any more than the equilibrium of a system of pulleys has to do with market-clearing. It simply represents the state to which the system will tend to return after a disturbance.

However, as we have seen, unemployment often takes a very long time to return to its original level. And it is not true that, once unemployment has risen, inflation starts to fall and continues to do so until unemployment returns to its original level. For example, in the EC inflation fell sharply in the early 1980s while unemployment was rising, but stabilized in the later 1980s when unemployment was still high but beginning to fall. This suggests that inflationary pressure is reduced not only by a high *level* of unemployment but also by *increases* in unemployment. Thus, if unemployment is falling (even though it is still high), inflation may not fall at all.

It is easy to see why inflation is affected not only by the level of unemployment but also by whether unemployment is rising or falling. There are three main reasons. When unemployment is rising, people are losing their jobs, and when the employed 'insiders' bargain with their employers the fear of job loss induces wage restraint. But when unemployment is falling, very few workers need worry about their jobs and the fall in unemployment fuels wage pressure.

Second, if unemployment is rising, this means that last year's unemployment was low relative to this year's. Thus, the unemployed 'outsiders' include no large backlog of long-term unemployed, and employers perceive the majority of the unemployed as employable. This helps to restrain inflationary pressure. By contrast, if last year's unemployment was high relative to this year's, there will be a backlog of long-term unemployed, who will have become demoralized and deskilled and will not be perceived as desirable by employers. In this situation a given amount of unemployment will be less effective in restraining inflation. Third, even if unemployment is high, when it is falling and the economy is expanding, costs tend to be increasing sharply as firms move towards full capacity with overtime increases, shift premia, and so on. This tends to put upward pressure on prices and hence wages which will be exacerbated if employment adjustment or turnover costs make it expensive to expand employment.

If wage pressure depends not only on this year's level of unemployment but also on changes from last year's, we have to augment our concept of equilibrium unemployment. There is indeed a long-run equilibrium at which both unemployment and inflation will be stable.[3] We shall call this the long-run NAIRU (non-accelerating-inflation rate of unemployment).[4] But if last year we were above the long-run NAIRU and then fell back to it immediately, we would have rising inflation. There is however some 'short-run NAIRU', which *would* be consistent with stable inflation, and which of course depends on last year's unemployment. In this view of the world there is short-term 'hysteresis', in the sense that past events affect the current short-run

NAIRU. But there is no long-term 'hysteresis': there is a unique long-run NAIRU. In the end, the unemployment rate always reverts. And employment always adjusts to the size of the labour force.

Thus, the theory of unemployment goes as follows.

1. There is a long-run NAIRU which depends on social and economic variables. It is of course subject to long-term change (e.g. from different benefit systems or wage-bargaining arrangements) and to temporary change (e.g. from changes in oil prices).
2. Demand shocks move employment away from the NAIRU and move inflation in the same direction as employment.
3. Supply shocks move employment by moving the NAIRU, and move inflation in the opposite direction to employment.
4. Once unemployment is away from the NAIRU, it takes some time to return even if inflation is stable.

There is no point trying to label this theory as Keynesian or classical. It has classical elements (the NAIRU) and it has Keynesian elements (the role of demand and the role of persistence). So it is best to avoid the use of those terms, which mean something different to every reader.

The issue of market-clearing

As we have said, our concept of equilibrium does not imply market-clearing. Everyone knows that some people fail to get jobs, while others who are just like them succeed. What explains this process of job rationing? Why do firms not drop their wages, so that it becomes worthwhile for them to employ the extra workers? There are two main explanations. First, every personnel director will tell his board that this will reduce morale and cause trained workers to quit; the losses from this would outweigh the savings made on the lower wages. This is the 'efficiency wage' explanation. Second, the union may prevent the firm paying less.

But we have to be careful here. Even when unemployment is high, there are not queues for all vacancies. There is a small secondary sector of the labour market that does more or less clear (e.g. in catering, cleaning, some maintenance and repairs, retailing and construction). If people are unemployed, it is generally because they have decided against these jobs. They are however willing to work in a range of good 'primary' sector jobs, but they cannot get them. In this sense unemployment is both voluntary and involuntary.

Outline

Our task now is to develop this framework in more detail, and to use it to explain the facts. When we have done this, we should have a good idea about how unemployment can be reduced.

Thus we shall proceed by asking (and answering) the following ten questions:

- What determines equilibrium unemployment?
- Why does unemployment fluctuate?
- How do real wages relate to unemployment?
- If labour markets don't clear, why don't wages fall?
- How do import prices, taxes, and productivity affect unemployment?
- How does job-search behaviour affect unemployment?
- Is unemployment voluntary or involuntary?
- Why are some groups more unemployed than others?
- Why has unemployment differed between countries?
- How can unemployment be reduced?

3

What Determines Equilibrium Unemployment?

When buoyant demand reduces unemployment (at least, relative to recent average values) inflationary pressure develops, and when unemployment is high the reverse happens. In a sequence of years when the average price level is stable, inflationary pressure means rising prices; in a sequence of years when the average inflation rate is stable, inflationary pressure means rising inflation. Since around 1970 the latter case is the most relevant, and, as we have seen in Fig. 5(c), there is a clear positive relation between changes in inflation and the (detrended) employment rate. In each case high employment applies an impulse to the inertial process by which prices are evolving, and a wage–price spiral develops with wages and prices chasing each other upwards.

Wage-setting and price-setting

Why exactly does a wage–price spiral develop? The answer is that stable inflation requires consistency between

(a) the way in which wage-setters set wages (W) relative to prices (P), and
(b) the way in which price-setters set prices (P) relative to wages (W).

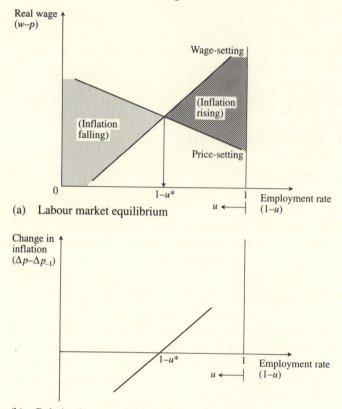

(a)   Labour market equilibrium

(b)   Relation between unemployment and inflation

Fig. 6. *Unemployment and inflation.*

Only if the real wage (W/P) desired by wage-setters is the same as that desired by price-setters will inflation be stable. *And the variable which brings about this consistency is the level of unemployment.* This affects the wage mark-up and also (probably) the price mark-up. Thus, inflation will be stable only if unemployment is at the appropriate equilibrium level. By the same token, if financial policy ensures that inflation *is* stable, then unemployment will adjust to its equilibrium level in the long run.

Thus, unemployment is the mechanism which eventually ensures that the claims on the national output are compatible. If a worker produces 100 units of output priced at \$1 and wage-setters set his wage at \$60, then the worker gets 60 units of output and profit-receivers get 40 units per worker. If this is what wage-setters and price-setters intended, we have an equilibrium. But if wage-setters aim at 61 units ($W/P = 61$) and price-setters aim to provide profits per worker equal to 41 units ($W/P = 59$), we have an inconsistency. This leads to a wage–price spiral, as wage-setters try to recoup the losses imposed on them by price-setters, and vice versa. In the long run, unemployment will have to be higher in order to reduce both sets of claims until they are equal with each other. Only in this way is the wage–price spiral eliminated.

There is another equally important spiral which unemployment eliminates. This is the wage–wage spiral. If unemployment is too low, wage-setters will try to raise their relative wage. Only if the labour market is slack enough will this leapfrogging be eliminated. In equilibrium, unemployment must be high enough to induce each particular wage-bargain to equal the bargain expected to prevail elsewhere.

We can illustrate all this with the following stripped-down model, in which parameter symbols are written as positive. We look first at price-setting, then at wage-setting. Prices (of value added) are set as a mark-up on expected wages. The mark-up tends to rise with the level of activity although this effect may not be very strong. (And if it is non-existent we have 'normal-cost' pricing.) Thus,

$$p - w^e = \beta_0 - \beta_1 u \quad (\beta_1 \geqslant 0), \tag{1}$$

where p is log prices, w^e log expected wages, and u the unemployment rate. This is graphed in Fig. 6(a) as the intended real wage set by price-setting. It can, if one likes, be thought of as the 'feasible' real wage—that real wage which (for given productivity) price-setters are willing to concede.

We turn now to wage-setting. Wages are set as a mark-

up on expected prices, with the mark-up tending to rise as
the employment rate rises and unemployment falls. Hence

$$w - p^e = \gamma_0 - \gamma_1 u \qquad (\gamma_1 > 0). \tag{2}$$

This is graphed in Fig. 6(*a*) as the intended real wage set by
wage-setting. It is, if you like, the 'target' real wage which
wage-setters intend.

If actual wages and prices are at their 'expected' values
($p = p^e$, $w = w^e$), the equilibrium unemployment rate is
given by adding (1) and (2) to obtain

$$u^* = \frac{\beta_0 + \gamma_0}{\beta_1 + \gamma_1} \tag{3}$$

This is illustrated in Fig. 6(*a*). The wage-setting and price-
setting lines are drawn for $p - p^e = w - w^e = 0$, and their
intersection determines equilibrium unemployment and real
wages. Any factor that exogenously raises wage push (γ_0)
or price push (β_0) raises the equilibrium rate. Any factor
that raises real wage flexibility (γ_1) or price flexibility (β_1)
reduces the equilibrium rate.

Unemployment and changes in inflation

If expected values of prices and wages are *not* realized, we
have

$$u = \frac{\beta_0 + \gamma_0 - (p - p^e) - (w - w^e)}{\beta_1 + \gamma_1}$$

or

$$u = u^* - \frac{(p - p^e) + (w - w^e)}{\beta_1 + \gamma_1}$$

Assuming that the 'surprises' on wages and prices are similar,

$$u - u^* = - \frac{1}{\theta_1} (p - p^e), \tag{3'}$$

where $\theta_1 = (\beta_1 + \gamma_1)/2$, which is a measure of real wage and price flexibility. Thus, low unemployment is associated with positive price surprises.

Suppose that we are in a period when inflation (Δp) is not expected to change. Then it is perceived as a random walk with

$$\Delta p = \Delta p_{-1} + \epsilon,$$

where ϵ is white noise, Δ means the change since the previous period, and -1 means one period earlier. Then the rational forecast of inflation is

$$p^e - p_{-1} = \Delta p_{-1}.$$

In consequence, the price 'surprise', $p - p^e$, is

$$p - p^e = p - p_{-1} - \Delta p_{-1} = \Delta p - \Delta p_{-1} = \text{change in inflation.}$$

Price surprises are equivalent to increases in inflation. The same is true of wages.

Thus, equation (3') implies that

$$\Delta p - \Delta p_{-1} = - \theta_1 (u - u^*). \tag{3''}$$

This is a standard Phillips curve relation and is shown in Fig. 6(b). When unemployment is lower than u^*, inflation is increasing; and vice versa. Thus u^* can be thought of as the non-accelerating inflation rate of unemployment (NAIRU).

Notice that inflation in one year is influenced by previous inflation. There is thus an element of 'nominal inertia' in the system: nominal prices are influenced by past history, and not only by forces at work today. The explanation of nominal inertia which we have just given is that the past influences expectations (so that unemployment can fluctuate only if expectations turn out to be wrong). But in fact, as we shall explain later, nominal inertia arises also from staggered wage- and price-setting, and from the cost of changing wages and prices.

4
Why Does Unemployment Fluctuate?

In the long run, unemployment is determined entirely by long-run supply factors and equals the NAIRU (u^*). But in the short run, unemployment is determined by the interaction of aggregate demand and short-run aggregate supply. Short-run aggregate supply is given by

$$\Delta p - \Delta p_{-1} = -\theta_1 (u - u^*). \qquad (3'')$$

Aggregate demand is (with suitable choice of units) given by

$$u = -\frac{1}{\lambda}(m - p),$$

where m is the log of nominal GDP (adjusted for trend real growth). This aggregate demand relation implies that

$$\Delta p = \Delta m + \lambda(u - u_{-1}). \qquad (4)$$

This demand curve (D) is drawn together with the short-run aggregate supply curve (SRS) in Fig. 7(a). Together they determine the current levels of unemployment and inflation.

This framework brings out two key points. First, a 'demand shock', associated with a rise in Δm, will shift D outwards and thus raise both inflation and employment. By contrast, a 'supply shock', associated with a rise in u^*, will also raise inflation but will reduce employment (see Fig. 7(b)).

The general expression for the level of unemployment is, from combining (3") and (4),

$$u = \frac{1}{\theta_1 + \lambda} \left[\theta_1 u^* + \lambda u_{-1} - (\Delta m - \Delta p_{-1}) \right] \tag{5}$$

If Δm is constant for long enough and u^* is constant, Δp_{-1} converges on Δm and unemployment converges on u^*.

This provides an adequate framework for analysing the history of our times.

Demand and supply shocks

In the late 1960s and early 1970s demand was stoked up, partly because of the Vietnam War. Δm rose, dragging up inflation and driving down unemployment (see Fig. 7(a)).

But, partly as a result, there followed in 1973–4 the huge rise in the price of commodities, including oil, mostly supplied from outside the OECD. This (together with greater union militancy) raised the NAIRU in the OECD, at least for a time—from u_0^* to u_1^*. The short-run aggregate supply curve therefore shifted leftward (see Fig. 7(b)). In consequence inflation rose, but this time unemployment rose also. Many analysts at the time were baffled, since they had been educated to believe that inflation and unemployment always moved in opposite directions, as in Fig. 6(b). But this is true only when the shock is a demand shock: with a supply shock, inflation and unemployment move together.

Following on the first oil shock there was little demand deflation, so inflation fell little. But then followed the second oil shock in 1979–80, which again raised inflation and also unemployment. At this point the electors in most countries declared that enough was enough: inflation must be reduced. There followed massive demand deflation in all countries, and by 1985 OECD inflation had been reduced to the same level as existed in 1969. At the same time, OECD unemployment had risen by over a half.

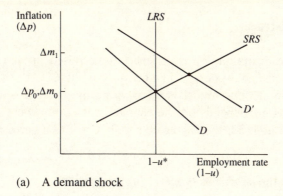

(a)   A demand shock

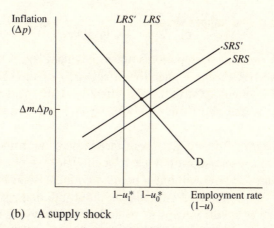

(b)   A supply shock

Fig. 7. *Aggregate supply and demand:* (a) *a demand shock;* (b) *a supply shock.*

It was already evident by the middle 1980s that European inflation was coming down much more slowly than might have been expected, given the high level of unemployment. Since then performance has been even more disappointing. In 1985–6 we had a beneficial oil shock, and real commodity prices fell to the same level as around 1960. One would have expected this to produce a major fall in inflation and unemployment. There has indeed been some fall in unemployment, but OECD inflation by the end of the 1980s was the same as in 1985.

Persistence

These disappointing experiences have raised in sharp form
the issue of hysteresis in unemployment. Clearly, we have
to modify our model to allow for this. If wage and price
behaviour depends on the change in unemployment as well
as on its level, the aggregate supply curve (3") becomes[5]

$$\Delta p = \Delta p_{-1} - \theta_1(u - u^*) - \theta_{11}(u - u_{-1}). \tag{3'''}$$

Thus the short-run NAIRU (u_s^*) is given by

$$u_s^* = \frac{\theta_1}{\theta_1 + \theta_{11}} u^* + \frac{\theta_{11}}{\theta_1 + \theta_{11}} u_{-1}.$$

It lies between last period's unemployment and the long-
run NAIRU.[6] The higher the effect (θ_{11}) of the change in
unemployment relative to the effect (θ_1) of the level, the
nearer is the short-run NAIRU to last year's unemployment.

In terms of policy, hysteresis means that, once unemployment has risen, it cannot be brought back at once to the
long-run NAIRU without a permanent increase in inflation.
But it can be reduced gradually without inflation rising.

Hysteresis clearly helps us to understand why inflation did
not fall in Europe over the later 1980s. But there is also
another important element: the fact that extra unemployment has a smaller effect on wages when unemployment is
already high than it does in a tighter labour market. One
can think of many reasons why wage-setters would respond
in this non-linear way. For example, if an employer found he
had 2 applicants per job rather than 1, he would relax his
wage by more than if he had 12 applicants rather than 11.

Thus, the large extra unemployment of the 1980s had
quite a small deflationary effect. By contrast, the small
excess demand of the early 1970s produced quite large
increases in inflation.

5

How do Real Wages Relate to Unemployment?

The next question is, Where do real wages fit into the picture? It is often claimed that unemployment occurs because real wages are too high. Is this true? We can discuss the issue first in the long term and then in the short term. Both analyses draw on Fig. 6(*a*).

In the *long term* the issue is whether the mark-up of price over wage-cost rises with the level of economic activity; i.e., does the 'price-setting' real wage in Fig. 6(*a*) slope down to the right? This is a matter of controversy.

We can begin with the extreme case of 'normal-cost pricing', where (for a given level of trend productivity) the price mark-up is constant. In this case, if there were a spontaneous increase in wage pressure, it would not actually have any effect on real wages. But it would raise unemployment, as illustrated in Fig. 8. Thus the problem is not that real wages are too high, but that too high real wages are desired at given unemployment. This is always the root of the problem. There can be extreme problems of wage pressure without any evidence of an actual 'wage gap', and indeed the whole concept of the wage gap tends to confuse rather than clarify.

Of course, if economic activity does raise the price mark-up, as in Fig. 6(*a*), then extra wage pressure will indeed raise real wages as well as unemployment. But the ultimate cause of both unemployment and higher real wages is the

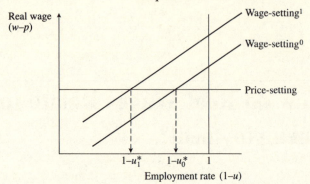

Fig. 8. *With normal-cost pricing, real wage pressure raises unemployment but not real wages.*

wage pressure. Unemployment and real wages are jointly determined.

Turning to the *short term*, what happens to the real wage if a demand shock reduces unemployment? This depends on the structure of wage- and price-setting. To understand what is going on, it is helpful to rewrite the price equation as

$$p - w = \beta_0 - \beta_1 u - (w - w^e) \qquad (1')$$

and the wage equation as

$$w - p = \gamma_0 - \gamma_1 u - (p - p^e). \qquad (2')$$

It can be seen that, when inflation is increasing ($p - p^e > 0$, $w - w^e > 0$), the real wage is below what wage-setters intended, and above what price-setters intended. Thus, inflation is the mechanism that reconciles the struggle for shares of the national cake, by cheating both price-setters (capitalists) and wage-setters of what they intended. Hence the observed point is in the darkly shaded area in Fig. 6(*a*).[7]

But has the real wage moved up or down, compared with its equilibrium value? It depends. With equal inertia in both price- and wage-setting (i.e. $p - p^e = w - w^e$), the real wage rises so long as the wage-setting line is steeper than the price-setting line.[8] However, in practice there may be different degrees of wage and price inertia, and in most countries there appears to be more inertia in the formation of prices than wages. As a consequence p tends to be closer to p^e than w is to w^e, making it likely that observations will be close to the wage setting line leading to real wages rising in a demand boom. Putting this another way, if prices are stickier than wages, when demand increases, wages respond more rapidly than prices and real wages go up.

In fact, in our estimates real wages generally rise in a (demand-led) boom. The same is true of most of the standard macroeconomic models used in public debate. This reflects the fact that our model and theirs are very similar. So is the model of the man-in-the-street, who also believes that real wages rise in a boom. However, this is not what many economic textbooks say.

Relation to new classical macroeconomics

We should at this point make clear how our approach compares with the 'new classical macroeconomics' of Lucas (1972) and others. Both approaches have an equilibrium unemployment rate. But in the new classical model, what we call the 'price equation' is thought of as the labour demand curve of firms selling in perfectly competitive product markets. Labour demand is unaffected by money illusion or nominal inertia. It follows that real wages fall in demand-led booms. In our own approach we prefer to think of the 'price equation' as representing a locus of price–employment combinations consistent with profit-maximizing behaviour by monopolistically competitive firms. There is substantial nominal inertia in pricing largely because of staggered price-setting and the costs of frequent price adjustment.

Turning to the 'wage equation', the new classical model would call this a labour supply equation, which is held to be relevant since the labour market is continuously in balance, with workers on their supply curves. In booms, workers underestimate prices and think real wages have risen when they have not. This elicits an increased labour supply. There is however no satisfactory evidence to support the view that cyclical changes in employment correspond to changes in the amount of labour people wish to supply. As Annex 1 shows, the 'intertemporal substitution' theory of fluctuations has little factual basis.

By contrast, we think of the relationship not as a supply equation but as a wage-setting equation, with wages tending to exceed the supply price of labour. If wages are set unilaterally by firms, it may still be in their interest to set them above the market-clearing level ('efficiency wages'). Or firms may be forced to do this by union bargaining.

So labour-market-clearing is not a necessary condition for equilibrium. Market-clearing is even less likely during fluctuations. Disequilibrium wage-setting does not necessarily require mis-perception by workers, since nominal inertia can be readily explained by overlapping wage contracts.

Although our interpretation of the structural model differs so sharply from the new classical model, it remains true that the reduced forms are indistinguishable. Equation (3') is the so-called 'Lucas supply curve'. In Lucas's interpretation, the price surprise causes the difference in output; in ours (as in the original Phillips curve) the difference in output causes the price disturbance. But the relationship is the same.

Since the policy implications of the two approaches are so different, it is very desirable to find independent evidence (from structural relationships, surveys, and the like) that will enable us to distinguish between the two descriptions of the world. We shall be doing this repeatedly.

Another rival analysis of fluctuations is the theory of equilibrium real business cycles. This is based on the idea that fluctuations originate mainly not from demand shocks but from exogenous supply shocks to the level of productivity. This approach does have the merit of predicting that real wages rise when output rises. But there are good reasons to think that most cyclical productivity fluctuations are not exogenous but are due to differential labour hoarding or work effort over the cycle (see Burnside *et al.* (1993) for evidence in favour). And, worst of all, the theory implies that inflation falls when output rises, despite the clear evidence of episodes when demand pressure increased both output and inflation.

6

If Labour Markets Don't Clear, Why Don't Wages Fall?

Our approach, therefore, is one in which some labour markets generally fail to clear, and jobs are rationed. The main evidence for this are the queues of applicants for many jobs. Though queues exist for jobs at all levels, skilled workers can generally get jobs in less skilled work. Thus, less skilled workers account for most of the unemployment problem. So why is there this persistent rationing of jobs?

It is not, as in some 'disequilibrium' models, because firms cannot sell all they wish on the product market. It is quite true that they cannot, but this would be so even if there were full employment. For most firms have some monopoly power and therefore set prices above marginal cost in order to maximize profit. This means that they would be happy to sell more at the price they have set, if anyone would buy it. But, unlike the unfortunate workers, who did not set the price of their labour, the suppliers of goods did set the price. The firms are therefore rationed in a quite different sense. We thus reject the goods-market rationing models of Barro and Grossman (1971) and Malinvaud (1977), who assume an arbitrarily rigid price which prevents perfect competitors from selling all they want to: we know of no mechanism that could sustain such a price.

The key problem is in the labour market and revolves around the issue of what stops wages falling when there is an excess supply of labour. There can be two classes of explanation. Either firms are not free to choose the wage, and wage bargaining forces them to pay more than they wish; or, if firms are free to choose and still pay more than the supply price of labour, it must be in their interest to do so.

Efficiency wages

We begin with the case where firms freely choose to pay a high wage in order to maintain high morale and encourage effort. It is easy to see how behaviour of this kind could lead to a pattern of wage-setting where wages are higher than the minimum at which people are willing to work— hence the queue for jobs.

Suppose that in firm i effort per worker is given by

$$E_i = e\left(\frac{W_i}{W^e}, \; u\right) \quad (e_1, e_2 > 0; \; e_{11}, e_{12} < 0),$$
$$\qquad\qquad + \qquad +$$

where W_i is the wage in the firm and W^e the expected prevailing wage outside. High relative wages elicit effort. So does high unemployment (though high unemployment diminishes the marginal effect of financial reward). Unemployment has this effect both because it affects the ease of finding another job if you lose this one, and because it affects the ease of shifting from one job to another without positive support from the current employer.

The representative firm now chooses W_i and P_i to maximize profit, which is

$$\Pi_i = R(E_i N_i) - W_i N_i = R(E_i N_i) - \frac{W_i}{E_i} E_i N_i,$$

where $R(\cdot)$ is revenue and N_i is employment. The firm will chose W_i to minimize cost per unit of effort (W_i/E_i), and then will choose N_i to maximize profit. The firm will always find it worthwhile to raise the wage so long as a 1 per cent rise in wages brings forth a more than 1 per cent rise in effort. But, once this ceases to be the case, the firm will stop raising wages. Thus, the optimum wage is where the elasticity of effort with respect to the wage is unity.

This can be seen clearly from the condition for choosing W_i to maximize E_i/W_i, which requires

$$W_i \frac{\partial E_i}{\partial W_i} - E_i = 0.$$

Hence

$$e_1\left(\frac{W_i}{W^e}, u\right) \frac{W_i}{W^e} = e\left(\frac{W_i}{W^e}, u\right),$$

where e_1 is the derivative with respect to the expected relative wage.

We now come to the crucial point about general equilibrium. If inflation is stable, the representative firm must willingly set its wage equal to the expected prevailing wage. There must be no leapfrogging and no wage–wage spiral. Thus, $W_i = W^e$ is *the* condition for equilibrium. And it is unemployment that brings this about. So equilibrium unemployment (u^*) is given by

$$e_1(1, u^*) = e(1, u^*).$$

It is helpful now to relate this to the system of price and wage equations shown in Fig. 6(a). For an individual firm facing a given demand curve, the choice of price is equivalent to the choice of employment. Thus, maximizing profit with respect to employment gives the price equation. This requires

$$\frac{\partial \Pi_i}{\partial N_i} = R'_i E_i - W_i = 0.$$

Suppose for simplicity that output equals $E_i N_i$. Then $R'_i = P_i(1 - 1/\eta)$, where η is the elasticity of demand facing the individual firm. It is convenient to write $(1 - 1/\eta)$ as κ, which is a measure of product market competitiveness,

whose maximum value is unity. Hence the real product
wage determined by price-setting is

$$\frac{W_i}{P_i} = \kappa e \left(\frac{W_i}{W^e}, u\right).$$

$$+ \quad +$$

But in aggregate the price in the representative firm will
equal the price in all other firms ($P_i = P$), and similarly
with the wage ($W_i = W$). So the price equation is

$$\frac{W}{P} = \kappa e \left(\frac{W}{W^e}, u\right).$$

$$+ \quad +$$

As employment rises, real wages fall (as in Fig. 6(*a*)), but
they are increased by rising inflation. Similarly, maximizing
profit with respect to the wage leads to a general equilibrium wage equation

$$\frac{W}{P} = \kappa \frac{W}{W^e} e_1 \left(\frac{W}{W^e}, u\right).$$

$$- \quad -$$

As employment rises, real wages rise (again as in Fig. 6(*a*)).
  We have so far posed the firm's problem in terms of
morale and effort. But there are of course many other reasons why a firm can gain by raising its relative wages. In
particular, higher wages help it to retain and recruit workers. At the same time, higher unemployment raises profits
by reducing quits and vacancies. So the efficiency-wage
model reflects the whole range of well established person-
nel-management practices. And these practices lead firms to
offer wages above the market-clearing level.
  What evidence is there that this sort of thing is in fact
happening, i.e. that workers are receiving rents, even in the
absence of unions? One piece of evidence is the queues of
applicants. Other, more direct, evidence comes from the
fact that wages of otherwise identical workers differ widely

between firms and industries, and when individual workers move to 'high-wage' industries most of them get wage increases. The high-wage industries are mostly those where the morale of the workers matters more: they use valuable equipment, or their performance is more difficult to monitor. In this case we have an effort function which includes a firm-specific variable λ_i,

$$e\left(\frac{W_i}{W^e}, u, \lambda_i\right),$$

and the optimum wage for a firm is given by

$$e_1\left(\frac{W_i}{W^e}, u, \lambda_i\right)\frac{W_i}{W^e} = e\left(\frac{W_i}{W^e}, u, \lambda_i\right).$$

Thus, wages will be different in each firm. The law of one wage for each type of labour has been repealed. But we still have to prevent leapfrogging, and in equilibrium unemployment must be high enough to ensure that the average of the W_i does not exceed the prevailing expected wage level (W^e).

Unions and wage bargaining

In the USA, 'efficiency' considerations may well be the main source of non-market-clearing wages; after all, only around one-fifth of all workers are unionized, and fewer in the private sector. But in European countries there is no question that unions are important: in most European countries, over three-quarters of the workforce have wages that are covered by collective bargaining.

Unions have every incentive to set wages above market-clearing levels. And once again, unemployment has got to be high enough to stop leapfrogging between unions. We shall concentrate on the case where bargaining occurs in a decentralized way between each firm and its own union members, though essentially the same analysis would apply in the case of a bargain for a single industry.

In bargaining, the union's main concern is to push for higher wages. However, we cannot assume that this is their only concern, since in some cases a higher wage would lead to job losses for existing union employees.

Of course, not all job losses lead to layoffs, since they can sometimes be accommodated through the natural wastage which occurs when people quit. Thus, though in 1980 nearly half of all British workplaces cut the number of jobs, only 11 per cent of them had any compulsory redundancies (layoffs) and only 9 per cent any voluntary redundancies (with some overlap between the two).[9] The proportion of actual individuals who were made redundant was of course even lower than this—less than 5 per cent a year.

However, even if few lose their jobs, union wage policy will be restrained by fear of job loss if enough individuals *fear* that they will be unlucky. The more random the incidence of job loss, the more will wage-push be restrained by employment considerations. For simplicity, we shall assume that the workers to be laid off are selected randomly.

Apart from their anxieties about jobs, unions want higher wages. Firms, by contrast, push for lower wages. What determines the outcome? First, we must focus on the objectives of the union. Since layoffs are selected randomly, all workers obtain the same expected reward and so the union is simply concerned with the representative member. Her welfare depends on the excess of the real wage, W_i, over her expected alternative income outside the firm, A, multiplied by the chances that she will remain in the firm, S_i. So the union is concerned with $(W_i - A)S_i$, which may be thought of as the expected rent, or the expected additional reward which the representative union member will get from participation in the wage bargain. The firm, on the other hand, simply gets its profit, Π_i.

The wage outcome from the bargain is the one which maximizes the weighted sum of the log of the rewards to the union and the firm, namely

$$\max_{W_i} \; \beta \log (W_i - A)S_i + \log \Pi_i$$

where β can be thought of as the strength of the union in the bargain. This is the so-called Nash bargaining outcome which was originally justified axiomatically by Nash (1950, 1953) but can also be derived from strategic bargaining considerations (see Binmore *et al.*, 1986). It is important to recognize that the chances of survival in the firm, S_i, is decreasing in the wage as is the firm's profit. So the first order condition is

$$\frac{\beta}{(W_i - A)} + \frac{\beta}{S_i} \frac{\partial S_i}{\partial W_i} + \frac{1}{\Pi_i} \frac{\partial \Pi_i}{\partial W_i} = 0 \qquad (6)$$

By the envelope theorem, $\partial \Pi_i / \partial W_i = -N_i$ and hence the mark-up of the wage over outside opportunities is

$$\frac{W_i - A}{W_i} = (\epsilon_{SW} + W_i N_i / \beta \Pi_i)^{-1}$$

where $\epsilon_{SW} = - W_i \partial S_i / S_i \partial W_i$, the absolute elasticity of survival with respect to the wage.

In order to interpret this expression, we must look more closely at the terms on the right-hand side. If we suppose that the firm faces a product demand curve $Y_i = P_i^{-\eta}$ and has a production function $Y_i = N_i^\alpha K_i^{1-\alpha}$ (K_i is the firm's capital stock), then given a bargained wage W_i, the firm will maximize profit by setting employment N_i at

$$N_i / K_i = (W_i K_i^{1/\eta} / \alpha \kappa)^{-1/(1 - \alpha \kappa)}$$

where $\kappa = 1 - 1/\eta$, our measure of product market competition. It is then easy to show that the wage and profit shares are $\alpha \kappa$ and $1 - \alpha \kappa$ respectively and hence $W_i N_i / \Pi_i$ can be written as $\alpha \kappa / (1 - \alpha \kappa)$.

The survival function, S_i, reflects the probability that the average union member will remain employed. This is related to the wage via the fact that the number of jobs expected to be available depends on expected employment which, in its turn, depends inversely on the wage. So we

may write the absolute elasticity ϵ_{SW} as $\epsilon_{SN}\epsilon_{NW}$ and, from
the expression for employment (7), ϵ_{NW} is $1/(1 - \alpha\kappa)$. Using
this and the expression for labour's share, the mark-up of
the wage over outside opportunities given in equation (6)
becomes

$$\frac{W_i - A}{W_i} = \frac{1 - \alpha\kappa}{\epsilon_{SN} + \alpha\kappa/\beta} \qquad (8)$$

This is highly informative. It shows that the mark-up of the
wage over the outside alternative (A) is higher the higher is
union power (β), the lower is product market competitiveness (κ), and the lower is labour intensity (α). Thus, the
mark-up depends on the rents coming from product market
power and the quasi-rents coming from fixed capital—
together with the power of the union to appropriate these
rents.

Thus, when we look at wages in individual industries, we
are not surprised that these are higher (for given types of
worker) the more concentrated and the more capital-inten-
sive the industry. The mark-up is also higher if the average
worker is unlikely to lose his job if employment falls—for
example because of high natural wastage.

So much for the mark-up. The actual wage depends also
on the outside opportunities for disemployed workers (A).
Such workers have a chance of getting another job paying
the expected prevailing wage (W^e); if not, they will get
benefit (B). Their chance of getting a job is higher the less
unemployment there is, and we shall assume the chance is
$(1 - \varphi u)$. Thus, the expected outside income is

$$A = (1 - \varphi u)W^e + \varphi u B \quad (W^e > B).$$

Hence, since wages in a given firm are higher when outside
opportunities improve, wages will be higher the lower is
unemployment and the higher are benefits.

We can now look at the aggregate economy. If we have
an equilibrium with no leapfrogging, $W_i = W^e$. Unemployment adjusts to bring this about. In addition, the wage in

the representative firm is, by definition, equal to the aggregate wage: $W_i = W$. Thus, from the definition of A,

$$\frac{W_i - A}{W_i} = \varphi u \left(1 - \frac{B}{W} \right),$$

and hence, using (8),

$$u^* = \frac{1 - \alpha\kappa}{(\epsilon_{SN} + \alpha\kappa/\beta)(1 - B/W)\varphi}. \tag{8'}$$

If we take the real *level* of benefits as exogenous, this is a real wage equation, and equilibrium unemployment is found by combining it with a standard price equation. If (more realistically) we take the replacement *ratio* (B/W) as exogenous, (8') gives the direct expression for unemployment. It shows that unemployment is higher the greater the power of unions, the greater the rents from product market monopoly and fixed capital, and the higher the replacement ratio.

In this situation unemployment is involuntary. Wages have been set by a process which involves only the firm and its existing workers (the insiders). Provided the unions are strong enough, the resulting real wage exceeds the supply price of the unemployed outsiders.

Of course, if the firm could sack all its workers, this power would vanish. But the specific training embodied in the workforce makes this unprofitable, except in extreme circumstances. And two-tier wage structures, in which outsiders are hired at their supply price, are ruled out because insiders rightly fear that the extra low-wage workers would eventually dominate the union. So union bargaining leads to non-market-clearing wages and unemployment.

Clearly, if there were no rents, there would be no scope for union wage gains. It is therefore obvious that product markets in which there is easy entry for new firms are conducive to low unemployment.

Any union model of unemployment is (and always has been) a model of insider power. But a key issue that arises

in the light of recent experience is, Does insider power lead
to hysteresis?

Insider power as a source of hysteresis

The answer to the above question is by no means obvious.
For example, suppose that the jobs of the workers who
control the unions are in effect safe, regardless of feasible
variations in the wage. This could easily be the case if layoffs were in order of seniority. And, if so, employment considerations would have no effect on how hard the union
pushed for wages. In this case the NAIRU would be given
by (8') with ϵ_{SN} set equal to zero—since wages would have
no effect on the survival probability of the workers who
matter. The number of historically determined insiders
would have no effect on this period's NAIRU.

However, suppose layoffs are by random assignment. In
those firms in which there is a risk of layoffs, wages then
have a material effect on the relevant chances of survival
($\epsilon_{SN} > 0$). And if last year's workforce was large relative to
this year's expected employment, a higher wage this year
will certainly put existing workers' jobs at risk. Thus, ϵ_{SN}
will be large, wage pressure lower, and unemployment
lower. By contrast, if last year's workforce was small relative to this year's expected employment, then a higher wage
this year will involve little extra risk, since most existing
workers are already safe. Thus, ϵ_{SN} will be small, wage
pressure higher, and unemployment higher. If last year's
unemployment was high, therefore, this year's NAIRU will
be higher than it would be otherwise.

But how far does such insider power actually explain
hysteresis? It is not the main explanation. To investigate
this question we have to use micro-data and see how far
wages at the micro-level depend on lagged employment at
the micro-level—as opposed to lagged unemployment in the
outside labour market. Most studies that have examined
this issue have found that the time-series movements of

wages depend much more on the outside labour market
than on firm-specific, or even industry-specific, factors. It is
lagged unemployment in the outside labour market that
matters much more than lagged employment in the firm or
industry. Theory also suggests that the impact on wage-bargaining of changes in the number of insiders must be quite
small. So in order to find a proper explanation of why
lagged unemployment matters, we shall have to look at the
behaviour of the unemployed outsiders. We come to this in
Chapter 8.

Corporatism

Whether insider power generates hysteresis or not, it certainly increases the average level of unemployment. We
would therefore expect that countries with less insider
power would have lower unemployment (and perhaps lower
hysteresis). Which countries are those?

Insider power typically requires unions. But it also
requires a particular form of union organization—where
unions bargain with their employers on a firm-by-firm (or
possibly industry-by-industry) basis. In such a context the
unions know that, if their workers become disemployed or
go on strike, they have a good chance of a job in the rest
of the economy. But suppose a single union bargained with
a single employers' federation on behalf of the whole workforce: there is then no 'rest of the economy' on which the
workers can fall back. (Alternative expected income, A, is
zero, assuming that benefits are financed by taxes on
employed workers.) The unions are now much weaker in
the bargain. In fact, there is a good chance that under this
centralized system the bargained outcome will be consistent
with full employment. In addition, if the employers bargain
as a whole, they will have no efficiency-wage incentive to
bid up wages against each other.

We can illustrate the different union objectives in Fig. 9
and show how, if wages were set by unions (with no

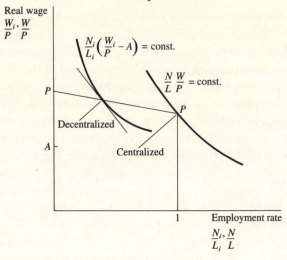

Fig. 9. *Centralized versus decentralized bargaining.*

employer resistance), a national union would be much more likely to choose full employment. The line PP indicates the 'feasible' real wage function of the economy as set by price-setting. This has a corner at full employment. Subject to this constraint, suppose the national union federation would like to maximize the expected income of each member of the labour force, i.e. the expected value of $(N/L)(W/P)$. So, unless the (absolute) elasticity of the real wage with respect to employment exceeds unity, which is most unlikely, the national union would choose the corner solution.

By comparison, a decentralized union can for simplicity be thought of as wishing to maximize the expected value of

$$\frac{N_i}{L_i}\left(\frac{W_i}{P} - A\right),$$

where L_i is the firm's 'share' of the labour force. This differs from the objective of the national union because there is a 'rest of the economy' to which disemployed workers can resort, which offers an expected income A. Thus, a decentralized union would never want a wage below *A*, and the isoquants for its objective function are thus asymptotic to *A*. (By contrast, the isoquants for the centralized union are much steeper and asymptotic to zero on the vertical axis.) The decentralized union is thus less likely to want a wage consistent with full employment.

And there is a further point. A union in a single firm can make its workers better off by raising their wage relative to other wages (and thus to the general price level). A national union cannot do this. It can only move back up the line *PP* in Fig. 9. The single representative firm too has to end up on the line *PP*, but it perceives its trade-off differently, with imperfect competition providing it with scope for additional real wage increases through increasing the relative price of its product. This is shown by the steeper line in Fig. 9— leading to a second source of extra unemployment when there is decentralized bargaining.

It is thus not surprising that the ordering of countries by unemployment is roughly as follows:

Unemployment	*Countries*	*Unions*
High	EC	Pervasive and decentralized
Medium	USA	Limited
Low	Scandinavia, Austria	Pervasive and centralized

Clearly, unemployment is also affected by factors others than the system of wage determination. The other key factor that affects unemployment is the behaviour of the unemployed themselves. We come to this in Chapters 8 and 9.

But first we need to ask how import prices, taxes, and productivity affect unemployment, using the framework we have just developed.

7

How do Import Prices, Taxes, and Productivity Affect Unemployment?

The answer is that, in the long run, they do not. If productivity, or living standards generally, had a long-run effect on unemployment, unemployment could not be untrended. And the theories we have been developing are consistent with this. Changes in taxes and import prices are in the long run borne by labour, with no change in unemployment. Similarly (at least with Cobb–Douglas production functions), productivity gains affect price- and wage-setting equally, with no change in equilibrium unemployment.

But in the short run things are very different. This is because the psychology of workers is more complicated than we have so far allowed for. Workers value not only the level of their real consumption wage, but also how it compares with what they expected it to be (or what they think is fair). For simplicity, we can think of people's expected living standards as a multiple of what they had last year. When external shocks like import price shocks, tax increases, or falls in productivity growth reduce the feasible growth of real consumption wages, this generates more wage pressure, which (in equilibrium) requires more unemployment to offset it.

We can illustrate this, first assuming efficiency wages and then wage bargaining. For simplicity, we now assume that effort is given by

$$e\left(\frac{R_i}{R_{i,-1}}, u\right) \qquad (e_1, e_2 > 0;\ e_{11}, e_{12} < 0),$$

where R_i is the real *consumption* wage and $R_{i,-1}$ is its lagged value. The real consumption wage is given by[10]

$$
\begin{aligned}
\log R &= \log(\text{net wage}) - \log(\text{consumer price}) \\
&= \log W - t_1 - t_2 - (\log P + t_3 + s_m \log P_m/P) + \text{const.} \\
&= \log W/P - (t_1 + t_2 + t_3 + s_m \log P_m/P) + \text{const.}
\end{aligned}
$$

Here W is labour cost, t_1 is the rate of labour taxes paid by the employer, t_2 is the tax rate paid by the worker, and t_3 is the indirect tax rate. P_m/P is the price of imports relative to GDP and s_m the share of imports in final output.

When import prices rise, this raises consumer prices relative to the GDP deflator (P). Thus there is a wedge ($t_1 + t_2 + t_3 + s_m \log P_m/P$) between the real product wage (W/P) and the real consumption wage (R). This wedge can be increased either by a tax increase or by a terms-of-trade shock, like an oil price rise. When this happens wage pressure increases, and in equilibrium unemployment must rise to contain it.

To check on this, we turn to the condition for the efficient wage (see Section 6). In aggregate (with $R_i = R$), this condition implies

$$e_1\left(\underset{-}{\frac{R}{R_{-1}}}, \underset{-}{u}\right) \frac{R}{R_{-1}} = e\left(\underset{+}{\frac{R}{R_{-1}}}, \underset{+}{u}\right).$$

Given the signs of the functions as indicated, a fall in R/R_{-1} must lead to a rise in u. Thus an increase in the wedge raises unemployment. Equally, if productivity growth falls, reducing R/R_{-1}, unemployment will also rise.

A similar story applies in the case of wage-bargaining. The union maximand now depends on $R_{i,-1}$ as well as $R_i -A$. As a consequence, a rise in the wedge will increase the bargained wage mark-up, and hence equilibrium unemployment.

To capture these effects, we need to modify our basic aggregate supply curve (3‴) to

$$\Delta^2 p = - \theta_1(u - u^*) - \theta_{11}(u - u_{-1}) + \mu\Delta\text{wedge}, \quad (3‴')$$

where $\Delta^2 p = \Delta p - \Delta p_{-1}$. This is quite a major modification. It means for example that a country can for a time improve its inflation–unemployment trade-off by appreciating its real exchange rate, i.e. by reducing P_m/P and hence reducing the wedge. Thus, as P_m/P falls, employment can rise with no inflationary take-off (as it did for example in the USA in 1983–5). In Fig. 10 we chart as SS the relation between non-inflationary employment and P/P_m—choosing to measure the relative price this way round in order to make clear that this relation is the non-inflationary supply curve of the economy. In other words, our supply of output (and jobs) increases as our relative price rises. By the same token, a real depreciation reduces non-inflationary employment—as the Germans regularly pointed out to the Americans in 1983–5.

Some have argued that this makes the NAIRU a useless concept, since in the short run employment can be increased without inflation rising, provided the value of the

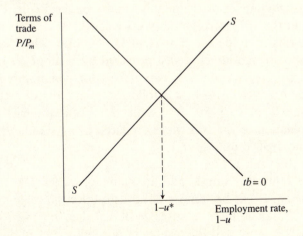

Fig. 10. *The terms of trade and the NAIRU.*

currency rises (e.g. through a fiscal expansion with a floating exchange rate). But this criticism misses the mark, for two reasons.

First, the implied loss of competitiveness will worsen the trade balance, and a trade deficit cannot be sustained indefinitely. For the trade balance relative to national income, tb, is given by

$$tb = a_0 - a_1(P/P_m) + a_2u,$$

and this shows that, if trade is to balance ($tb = 0$), a real appreciation in P/P_m will require a rise in unemployment—to restrain imports. This line for balanced trade is drawn in Fig. 10. Where the line crosses the SS line we have the long-run sustainable level of unemployment.

But there is a further point. The effect of the wedge on wage pressure probably does not last for ever. In the end, all changes in potential living standards are accepted by workers with no change in the NAIRU: it is only Δwedge that creates temporary changes in the NAIRU, and the wedge itself has no permanent effect. To this extent we never reach the long-run NAIRU shown in Fig. 10 because SS itself becomes vertical.

This framework of analysis is extremely helpful in looking at the impact of the supply shocks of the 1970s. First we had the 1973–4 commodity price shock which worsened the terms of trade for most OECD countries. This was a much more serious blow (as measured by $s_m\Delta\log P_m/P$) for most European countries than it was for commodity-rich North America. And it was followed by a blow of similar magnitude in 1979–80. For some European countries the combination of the two shocks reduced living standards by around 7 per cent.

At the same time, productivity growth slowed down throughout the world, and in many countries tax rates increased. It has been more difficult to trace the wage pressure effects of these changes in the wedge. But we shall show later how well the change in unemployment in different countries after each commodity price shock is explained by the size of that shock and by the institutional arrangements in the country.

As we have shown, real commodity prices eventually fell back in the mid-1980s to the same level as in the 1960s—since commodity prices tend to fall in world recessions (see Annex 2).[11] This in turn generated a world boom. But European unemployment was still high in the boom, since it had hardly recovered from the earlier recession. Why was this? As we have said, the hysteresis cannot be much explained by changes in the number of insiders. The clue lies in the behaviour of the outsiders.

8

How Does Job-Search Behaviour Affect Unemployment?

This is the element that has so far been lacking from our analysis. The unemployed have not appeared at all as people, whose behaviour matters—merely as pawns, whose number reconciles the claims to the national output. This is not the case, and it is time to expand the model to show how job search affects the equilibrium number of jobs.

The mechanism is this. Wage pressure builds up unless there is a sufficient excess supply of labour. (Firms bid up wages against each other and unions feel strong enough to press their claims.) But if the unemployed seek harder for jobs, this raises the effective excess supply of labour. (Firms can get workers more easily and disemployed people face fiercer competition for jobs.) Thus, if unemployed workers seek harder, there need be fewer of them in order to restrain wage pressure.

This leads us to modify our earlier wage equation to make wages depend on cu rather than u, where c measures the 'effectiveness' of the average unemployed job-seeker. To see exactly how this enters in, we shall start from a rather more structural wage function than we had originally. We shall now assume that, from the point of view of a worker facing possible unemployment, what matters is the chance of getting a job if he searches with a given effectiveness (say with $c = 1$). This chance is H/cU, where H is the number of

unemployed people hired per period, U is the number of
unemployed, and cU is the number of effective unemployed. But in equilibrium, the numbers hired equal the
numbers becoming unemployed. If the fraction of employed
workers (N) who become unemployed is s, this means that
in equilibrium $sN = H$, so that

$$\frac{H}{cU} = \frac{s}{cU/N} \simeq \frac{s}{cu}.$$

This becomes the relevant variable to explain the wage
pressure coming from the workers in wage bargaining.

There is also the wage pressure coming from the firms.
This depends on the chances of their filling each vacancy,
which is, in fact, uniquely related to H/cU.[12] So our new
wage equation is

$$w - p = \gamma_0 - \gamma_1(cu/s). \tag{2'}$$

In equilibrium, the more effective are the unemployed (i.e.
the higher is c), the lower is unemployment.

But how are we to measure c over time? One approach
might be to replace it by what determines it, including the
replacement ratio B/W. But there are many other factors
which also affect search effectiveness, including social attitudes to work, the stigma attaching to unemployment,
employers' attitudes, and so on. These are very difficult to
measure; but fortunately, there is some direct evidence on c
from the behaviour of unemployment in relation to vacancies.

The unemployment–vacancy relationship

Given the small amount of information economists have
about how their economies work, we need to exploit to the
utmost the information that vacancy data provide. In particular, we can obtain direct evidence on the effectiveness of
job search (c) by examining the movement of unemployment relative to the level of vacancies.

This is because there is a 'hiring' (or 'matching') function which explains the flow of unemployed people into work. This flow (H) depends positively on the number of vacancies (V) and also on the number of effective job-seekers (cU):

$$H = h(V, cU). \tag{9}$$

Provided the market is large enough, an equiproportional increase in vacancies and in effective job-seekers will induce an equiproportional increase in hirings. Hence the chances of finding a job are given by

$$\frac{H}{U} = ch\left(\frac{V}{cU}, 1\right). \tag{10}$$

Both (7) and (8) imply that, from knowledge of H, V, and U, we can infer changes in c.

Alternatively, we can use the fact that in equilibrium $H = sN$ to obtain the relationship

$$s = h\left(\frac{V}{N}, \frac{cU}{N}\right). \tag{10'}$$

This is the famous Beveridge curve (or U/V curve). For given s, shifts in this curve reflect shifts in c. As Fig. 11 shows, there has been a considerable increase in many European countries in the level of unemployment at given vacancies. (In the USA it shifted out but has now shifted back.) What could account for the outward shift of the European Beveridge curve?

From what we have said so far, the explanation would have to be a fall in search effectiveness (c) among the unemployed. However, there is another possible explanation to be considered. There could have been an increase in 'mismatch' between the pattern of unemployment and vacancies across sectors (i.e. regions, industries, or skill-groups). An increase in mismatch would shift out the U/V curve; for, provided the relation (10') between U/N and V/N in each sector is the same and convex to the origin, the

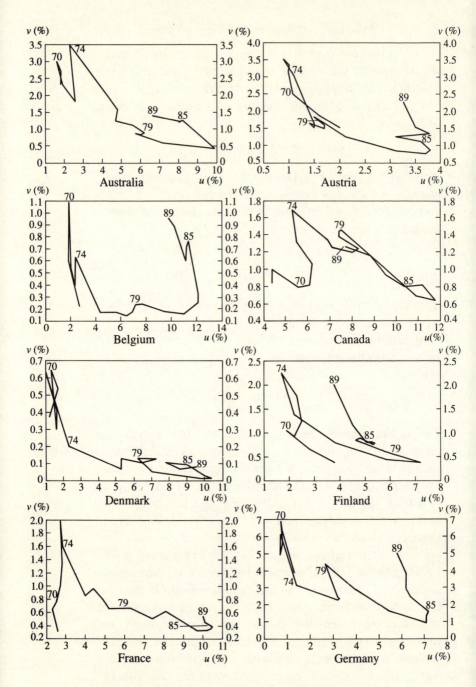

Fig. 11. *Vacancy rates* (v) *and unemployment rates* (u).
Source: Jackman *et al.* (1990).

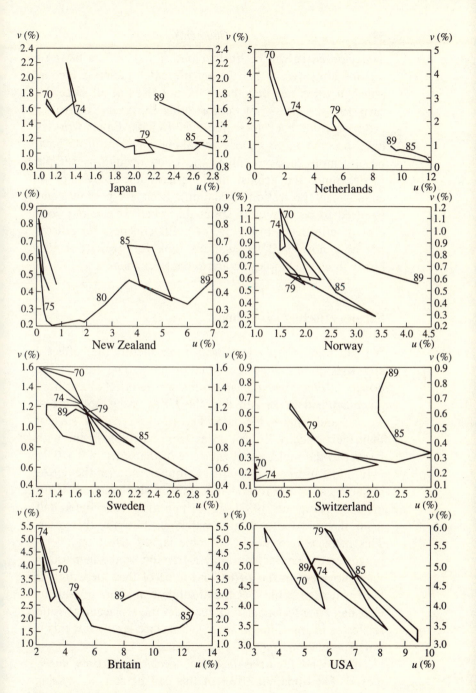

Fig. 11. (*cont.*)

aggregate curve will be 'further out' if U/V differs between sectors. However, if the relevant mismatch indices are computed, it turns out that they have not risen at all since the early 1970s in Britain or in most other European countries.

So we come back to search effectiveness. Either workers have become more choosy in taking jobs, or firms have become more choosy in filling vacancies (owing for example to discrimination against the long-term unemployed or to employment protection legislation). Both are possible, and we need to be very clear about this when we use our concept of effective job-seekers (cU). Effectiveness (c) reflects not only how hard the workers look for work, but also how willing the employers are to consider them.

Factors affecting job-finding

We can now identify two measurable factors that affect c and thus job-finding—see equation (10). The first is the *benefit–income (replacement) ratio*, whose effects have been much studied in Britain and the USA, using both cross-section and time-series data. The results typically suggest that the elasticity of exit rates from unemployment with respect to the replacement ratio are of the order of 0.2–0.9.

In most European countries, though not in the USA, replacement ratios rose significantly in the 1960s or 1970s or both (Emerson 1988). In Britain they rose by a half from the mid-1950s to the mid-1960s but not thereafter. This increase may have had some lagged effect on unemployment but can explain only a fraction of the increase in unemployment in the 1970s, and none of that thereafter. It is of course possible that the absolute real value of benefits also has an effect (e.g. that the relevant replacement ratio relates to incomes above some subsistence level). But this is pure speculation.

The second factor is *how long people have been unemployed*. The apparent effect of this can be seen in striking form by comparing the exit rates from unemployment of people with different durations. Fig. 12 shows this for both Britain and the USA. In the USA there are very few long-term unemployed. But in Britain, where there are many, the exit rates are much lower for the long-term unemployed. One reason for this must be that the more energetic job-seekers find jobs first, so that the long-term unemployed include a higher proportion of less energetic people. But time-series evidence makes it clear that another reason is the direct effect of unemployment duration upon a given individual. Long-term unemployment both demoralizes the individual and is also used by employers as a (biased) screening device. Thus, if the average duration of unemployment rises, we can expect the average level of c to fall. Hence unemployment will rise relative to vacancies.

The exact degree to which duration affects exit rates cannot easily be resolved from studies of individual data, owing to the problem of unobserved differences between individuals. But aggregate time-series equations indicate a considerable effect of duration structure upon average exit rates.

Moreover, in regressions of the unemployment rate on the vacancy rate and the proportion of long-term unemployed, the latter term has a significant positive effect. The same is true when the proportion of the long-term unemployed is included in a real wage equation: it increases wage pressure. In other words, the long-term unemployed are much less effective inflation-fighters, since they are not part of the effective labour supply.

Between 1979 and 1986, the proportion of unemployed who had been out of work for over a year rose from around 20 to around 40 per cent in Britain. Using the regression estimates, this in itself would explain one-third of the outward shift of the U/V curve. Similar findings apply in other major European countries.

It is noticeable in Fig. 13 that all the countries where long-term unemployment has escalated have unemployment benefits of some kind that are available for a very long period, rather than running out after 6 months (as in the

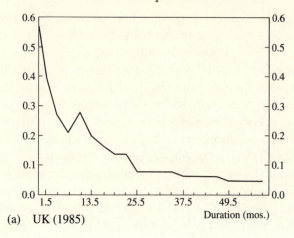

(a)   UK (1985)

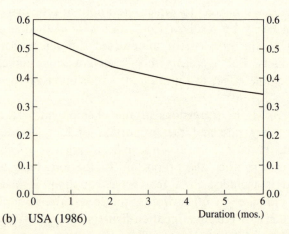

(b)   USA (1986)

Fig. 12. *Proportion of unemployed people leaving unemployment within the next 3 months, by existing duration of unemployment.*

Sources: (*a*) This is based on the outflow rates between April and July 1985 for those with the indicated durations in April. All but one of these rates come from taking the stock of unemployed for duration d in April and comparing it with those unemployed for duration $d + 1$ in July (or, where the stock data cover two quarters, $d + 2$ in October or, for four-quarter categories, $d + 4$ in the following April). The very first outflow rate, for those who just became unemployed, is com-

cont. opposite/

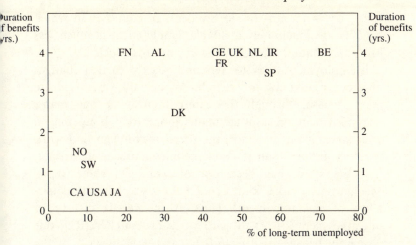

Fig. 13. *Maximum duration of benefit, 1985, and percentage of unemployed out of work for over a year, 1983–1988.*

Countries with indefinite benefits are graphed as having a 4-year duration.

AL: Australia; BE: Belgium; CA: Canada; DK: Denmark; FN: Finland; FR: France; GE: Germany; IR: Ireland; JA: Japan; NL: Netherlands; NO: Norway; SP: Spain; SW: Sweden; UK: United Kingdom; USA: United States.

Sources: proportion of long-term unemployed in total unemployed: OECD, *Employment Outlook*, July 1989 and July 1990, Table M; benefits: Table 5 below. Italy is omitted because it has no effective benefit system (see Annex 3, Table A1).

puted as

$$2 \left(1 - \frac{\text{stock of the unemployed under 3 mos. in Apr.}}{\text{inflow, Jan.–Apr.}} \right).$$

This is based on the assumption that the outflow rate over the first three months is constant, so that by the end of a quarter the remaining stock excludes one-half of those who leave within the first three months of their unemployment.
(*b*): OECD, *Employment Outlook*, Sept. 1988, Table 2.12.

USA) or 14 months (as in Sweden). In countries in which benefits are indefinitely available, employment is much less likely to rebound after a major downwards shock.

If employment does not rebound quickly, further changes may occur affecting job search. An unemployment culture may develop, through the external effect of one man's unemployment on another man's job search. If no one in your street is out of work, the social pressure to find work is much greater than if (as sometimes happens in Britain) half the street has been out of work for some years. Mechanisms of this kind could help to explain the persistence of unemployment. Thus, if the recent history of unemployment affects the current (short-run) NAIRU, this is mainly because it affects the search effectiveness of the unemployed 'outsiders', rather than because it reduces the number of 'insiders' in work.

9

Is Unemployment Voluntary or Involuntary?

In the last section we showed how the search behaviour of individuals affects equilibrium unemployment. At any moment there are outstanding vacancies as well as job-seekers, but it takes time to match them to each other. In consequence, unemployment and vacancies coexist. The harder people look for work, the lower unemployment will be, because wage pressure will be reduced (at any given level of unemployment).

This raises the question of whether unemployment is voluntary or involuntary. The question is fruitless. There are two aspects to reality:

1. There *is* job rationing, because individuals cannot just pick up a job.
2. The total number of jobs *does* respond to how hard people search.

To get a proper perspective on unemployment, it is essential to hold both points in view.

However, there are two further qualifications to be made to this picture:

3. For the semi- and un-skilled, there are often in fact very few well paid vacancies. Those that appear are snapped up overnight, and there are often hundreds of applicants who are, for all practical purposes, indistinguishable.

Employers report no shortage of labour to do these jobs. In Britain the proportion of employers in manufacturing who expect their output to be limited by shortages of non-skilled labour has averaged only 5 per cent over the last quarter-century (compared with 19 per cent for skilled workers). Thus, there are not many well-paid vacancies for less skilled labour in what we may call the 'primary sector'. Once people get these jobs, they tend not to quit.

4. However, though well-paid jobs are scarce, it is generally possible to find a badly paid one. For most of the unemployed (other than the handicapped) there is some vacancy they can pick up—in catering, cleaning, some retail stores, and small-scale repairs and maintenance. For those with sufficient enterprise, there is also self-employment. This whole sector we may call the 'secondary sector' (though in fact there is clearly a continuous spectrum of jobs). The secondary sector is market-clearing, in the sense that wages are not high enough to attract a queue of job-seekers, nor do vacancies last long since skill requirements are low. *In the secondary sector, if wages were lower employment would fall, because of reduced supply of labour; whereas in the primary sector, if wages were lower employment would grow, because of increased demand.*

Why are there people who would be willing to work in the primary sector but not in the secondary sector? It may be because it is harder to find a primary-sector job while already employed in the secondary sector than while unemployed. Another possible reason is that for some people life on unemployment income is preferable to life in the secondary sector. People vary in these respects, and for each person i there is some critical secondary-sector wage (W_i^*) at which they are just willing to work. The array of reservation wages (W_i^*) taken in ascending order provides the rising supply curve of labour to the secondary sector. Once the secondary-sector wage is determined, we know how many of those not employed in the primary sector will be employed in the secondary sector, and how many will be unemployed.

The primary and secondary sectors

We can illustrate the position in Fig. 14. The total labour force (employed plus unemployed) is L. We take this as exogenous, mainly on the grounds that the total labour force (male and female taken together) is not very responsive to changes in wages. All workers are willing to work in the primary sector. D_1 gives the demand relationship between primary employment and the primary-sector real wage (in units of general purchasing power). This wage is determined at the level shown, by the mechanism of efficiency wages or union bargaining already discussed. Thus, primary-sector employment is N_1. This leaves $L - N_1$ workers available for the secondary sector. We suppose that the distribution of reservation wages in this group is

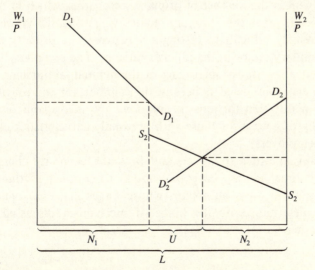

Fig. 14. *Unemployment in a two-sector model.*

independent of its size, with a minimum equal to the minimum height of S_2 and a maximum equal to its maximum height. D_2 gives the demand relationship between secondary employment and the secondary-sector real wage. In this sector the wage and employment are determined so that the market clears, with N_2 people being employed. This leaves $L - N_1 - N_2$ (= U) people unemployed.

These people are both involuntarily and voluntarily unemployed. They are willing to work in the primary sector at the going wage there, but have not so far found work; they are *not* willing to work in the secondary sector at the going wage there.

This account seems to capture the way most participants (firms and workers) perceive the equilibrium of the labour market. As time proceeds, some primary-sector firms expand, others contract. Thus, some people lose jobs in the primary sector and join those $L - N_1$ people outside who would like jobs in the primary sector. Some of these become unemployed, while others take secondary-sector jobs while continuing to look for better work.

How long those who become unemployed remain so depends on the general equilibrium of the system. The key element is the number of primary-sector jobs, which in turn depends on the primary-sector wage. It is however extremely difficult to distinguish between the primary and secondary sector in the official statistics. The secondary sector is also a fairly small part of the manual labour market. We have discussed it because it is important to recognize the reality that for most people *some* job is available (especially of course for those with personal characteristics liked by employers).

But in order to understand how the economy changes over time, it may be good enough to proceed as though there were only one sector, whose wages and employment are determined by the kinds of mechanism discussed in Chapter 6.[13]

10

Why are Some Groups More Unemployed Than Others?

We have proceeded so far as though all workers were the same, except that some of them have jobs and others haven't. But in fact, unemployment rates differ sharply between groups. Why is this, and how do these disparities affect the overall unemployment rate?

Causes of mismatch

As we have stressed before, unemployment mainly affects manual workers. Over three-quarters of unemployed men are manual workers. As Table 3 shows, this is mainly because they are more likely to *become* unemployed, not because they remain unemployed for much longer once unemployed.

Similarly, young people are more likely to be unemployed than older people, and this is again due to the fact that they are more likely to become unemployed. In fact, they remain unemployed for a rather shorter time on average than older people.

So here we have big differences in unemployment between occupations and age-groups which are persistent over time and across countries. They are not mainly related to general shifts in demand or supply, or to any resulting market disequilibrium. They are essentially equilibrium phenomena.

Table 3 *Unemployment by skill: flow and duration, Britain and USA*

	Britain (1984)			USA (1987)		
	Inflow rate (% per mo.)	Average duration (mos.)	Unemployment (%)	Inflow rate (% per mo.)	Average duration (mos.)	Unemployment (%)
Professional and managerial	0.50	11.2	5.3	0.74	3.0	2.3
Clerical	0.88	10.1	8.0 ⎫	1.58	2.6	4.3
Other non-manual	1.14	11.8	12.2 ⎭			
Skilled manual	1.02	14.2	12.6	1.97	2.9	6.1
Personal services ⎫	1.32	14.1	15.5	2.96	2.4	7.7
Other manual ⎭				2.84	3.0	9.4
All	0.94	12.8	10.8	2.23	2.6	6.22

Notes:

Inflow rate	= inflow/numbers employed
Outflow rate	= outflow/numbers unemployed
Unemployment rate	= numbers unemployed/numbers employed-or-unemployed

Monthly inflow and outflow are measured by numbers unemployed less than 1 month. In Britain the numbers in this category on the *Labour Force Survey* (*LFS*) definition of unemployment are only 70% of those in their first month of benefit receipt. The *General Household Survey* is broadly consistent with the *LFS*.

Source: Britain: *Labour Force Survey* tapes. This only records previous occupation and industry for those unemployed for under 3 years. The unemployment rate in each occupation is computed by taking the numbers unemployed for less than 3 years who were previously employed in the stated occupation and raising it by the ratio of total unemployed to numbers of unemployed reporting their previous occupation. A similar procedure is done for those unemployed for under 1 month. *USA*: *Employment and Earnings*, Jan. 1988, p. 175.

But they do result from *firm-level* disturbances, in which some firms are expanding and others are contracting. As firms are forced to contract, they lay off not their experienced non-manual staff, in whom they have sunk much firm-specific human capital, but their direct labour and to some extent those workers most recently hired (last-in, first-out). In addition, younger workers are more prone to quit. So what we are seeing is a stochastic equilibrium, involving a persistent mismatch between the pattern of the labour force (L) and the pattern of employment (N). How such a mismatch affects the overall unemployment rate we shall consider in a moment.

But first we turn to regional unemployment differences. Here we come nearer to something with a disequilibrium (i.e. transitional) element in it. Certainly these differences are related to shifts in labour demand, and to a failure of migration to keep pace. For example, we can see how the decline of textiles caused unemployment in New England in the 1960s and 1970s, while at the same time Texas boomed; and in the 1980s there was a complete reversal, as high tech boomed and oil faltered. Similarly, in Germany the North boomed in the 1960s and unemployment was relatively high in Bavaria; by the 1980s the decline of heavy industry had completely reversed the situation.

But in other countries regional unemployment differentials are much more persistent, with unemployment always higher in the North of England and the South of Italy. The differences here are sustained by steady one-way shifts in the pattern of demand, with migration never catching up. It is a steady-state disequilibrium.

The main reason why labour demand shifts from one region to another is that labour demand shifts between industries, and different regions are intensive in different industries. Thus, the degree of regional imbalance is related to the rate of change in industrial structure. In Britain regional unemployment differences were much greater in the interwar period than since the war—and so were the changes in industrial structure. If we compute the proportion of jobs in each industry in adjacent years and then take the changes in each proportion, we can sum the positive changes to get a measure of the proportion of employment 'changing industries'. This measure averaged 2.7 in 1924–39 and less than half as much (1.1) since 1950. The pattern for the USA is very similar (1.7 and 0.9). (The USA data relate to 1-digit and the British to 2-digit industries.)

But has turbulence increased since the 1960s in a way that could help to explain increased unemployment? The answer is a clear no. And for this reason, we are not surprised that the inflow into unemployment has not increased in most countries. The secular rise in unemployment is associated with increased duration and not with an increase in the rate of job loss.

Relation of mismatch to the NAIRU

The next issue is, How exactly do differences in unemployment rates affect average unemployment? In addressing this question, it is essential to banish the idea that the only interesting unemployment differences are those relating to disequilibrium problems of transitional adjustment. Even if age differences in unemployment reflect an equilibrium, it is still true that, by shifting labour demand from middle-aged to younger people, we could reduce the NAIRU. So we need a general framework for analysing the implications of unemployment disparities, from whatever source they may arise.

The basic idea is that, if there are unemployment disparities, this makes it more difficult to secure a low average level of unemployment. For the low-unemployment labour markets overheat while there is still high unemployment elsewhere. If, instead, we could increase unemployment by x per cent where it is low, and reduce it by x per cent where it is high, this would reduce wage pressure. For wages are more sensitive to unemployment when unemployment is low than when it is high.

The problem can be analysed quite simply within our standard framework, and ignoring nominal inertia. Assuming for simplicity normal-cost pricing ($\beta_1 = 0$), the price equation can be approximated by

$$p = \beta_0 + \sum_{i=1}^{I} \alpha_i w_i \quad (\sum \alpha_i = 1),$$

where there are I types of labour having (log) wages, w_i. There are separate wage equations for each type of labour, which evidence suggests have the concave form

$$w_i - p = \gamma_{0i} - \gamma_1 \log u_i \quad (i = 1, \ldots, I).$$

Substituting the wage equations into the price equation, and adding and subtracting $\gamma_1 \log u^*$, gives

$$\gamma_1 \left[\sum \alpha_i (\log u_i - \log u^*) + \log u^* \right] = \text{const.}$$

or[14]

$$\log u^* = -\sum \alpha_i \log(u_i / u^*) + \text{const.}$$

$$\simeq \frac{1}{2} \text{var} \frac{u_i}{u} + \text{const.}$$

Thus, the NAIRU depends on the variance of the relative unemployment rates. Hence equiproportional rises in unemployment rates do not increase the NAIRU. This is due to the curvature of the wage function, which empirically appears to be best represented by the double-log form.

We can now examine whether mismatch, measured in this way, has increased over time. Table 4 gives data for Britain on the variance of relative unemployment rates by occupation, age, region, and industry. There is no pattern of general increase since the mid-1970s. Similar results apply to regional and industrial patterns of unemployment in most of the main OECD countries. Though unemployment has risen, it has risen by much the same proportion in all groups—or, at least, its relative dispersion has not increased.

Turning to the shift in the U/V curve, this could be due to increased mismatch only if there were increased imbalance between the pattern of vacancies and unemployment. Imbalance of course exists, with vacancy rates low where unemployment rates are high. But there is no evidence in any major country that the misalignment has increased since the early 1970s.

This does not mean that mismatch is unimportant. If we add up the different mismatch indices in Table 4 for 1985 (treating them as independent), they have raised unemployment by some 40 per cent (half of the sum of the bottom row) above what it would otherwise be. Thus, when we come to policy, mismatch is a major issue. It is just that this is nothing new.

Table 4 *Variance of relative unemployment rates, Britain, 1974–1985 (%)*

	By occupation (5 groups)	By age (10 groups)	By travel-to-work area (322 areas)	By industry (10 groups)
1974	23	16	18	11
1975	14	19	22	13
1984	21	14	20	12
1985	22	23	24	14

Source: *General Household Survey*. Travel-to-work area data are available only for 1985 but have been inferred for other years, using regional data.

11
Why has Unemployment Differed Between Countries?

We are now ready to explain the differences in unemployment experience between countries. Table 1 shows the amazing spread of unemployment rates in 1990, and Table 5 gives similar data for 1983–8. Average unemployment was in 1990 roughly 9 per cent in the EC, 5½ per cent in the USA, 2 per cent in Japan, and 3 per cent in the EFTA countries. By contrast, in the 1960s the unemployment differences were small (in absolute terms): Britain, France, Germany, Belgium, the Netherlands, the EFTA countries, and Japan all had average unemployment below 2½ per cent; the USA had 4 per cent. Thus the challenge is to explain why the unemployment rates are now so different.

Static analysis

We shall begin with an extremely static approach to the issue, and then look more carefully at the different shocks that have affected different countries and how they have responded to them.

We began this chapter with the simplest possible Phillips curve (3"):

$$\Delta^2 p = -\theta_1(u - u^*).$$

Chapter 11

Table 5 *Unemployment experience of different countries, and treatment of the unemployed*

	(1) Unemployment rate % 1983–8	(2) % of long-term unemployed 1988	(3) Duration of unemployment benefit (yrs.) 1985	(4) Replacement ratio (%) 1985	(5) Expenditure on 'active' labour market programmes per unemployed person (as % of output per person) 1987
Belgium	11.3	78	Indef.	60	7.4
Denmark	9.0	29	2.5	90	7.9
France	9.9	45	3.75	57	3.9
Germany	6.7	47	Indef.	63	10.4
Ireland	16.4	66	Indef.	50	5.0
Italy	7.0	69	0.5	2	0.8
Netherlands	10.6	50	Indef.	70	2.7
Portugal	7.7	51	0.5	60	7.4
Spain	19.8	62	3.5	80	2.1
UK	10.7	45	Indef.	36	4.6
Australia	8.4	28	Indef.	39	2.8
New Zealand	4.6	—	Indef.	38	13.1
Canada	9.9	7	0.5	60	4.3
USA	7.1	7	0.5	50	2.4
Japan	2.7	21	0.5	60	5.6
Austria	3.6	13	Indef.	60	11.3
Finland	5.1	19	Indef.	75	12.9
Norway	2.7	6	1.5	65	9.8
Sweden	2.2	8	1.2	80	34.6
Switzerland	2.4	—	1.0	70	3.7

Sources: col. (1): see Annex 6, UK is UK(1); col. (2): OECD, *Employment Outlook*, July 1990, Tables M and P; cols. (3) and (4): mainly US Department of Health and Social Services, *Social Security Programs Throughout the World 1985 (Reserve Report No. 60)*; see also OECD, *Employment Outlook*, Sept. 1988, Tables 4.3 and 4.4. Further details in Annex 3; col. (5): OECD, *Employment Outlook*, Sept. 1988, Table 3.1.

If we let u^* depend on a vector of institutional variables z, we can rewrite this as

$$u = a_0 + a_1 z - \frac{1}{\theta_1}\Delta^2 p$$

or, for the ith country,

$$u_i = a_0 + a_1 z_i - a_2 \Delta^2 p_i. \qquad (11)$$

We can then attempt to explain the average unemployment rate (1983–8) in each country by its current institutional structures (z), and the degree of disinflation ($- \Delta^2 p$).

On the basis of our analysis so far, we would expect the NAIRU to be affected by the following variables in the manner shown:

	Effect
Duration of unemployment benefits	+
Replacement ratio	+
Expenditure on national manpower policies	−
Union coverage	+
Co-ordinated bargaining by unions	−
Co-ordinated bargaining by employers	−

So the first task is to look at the basic institutional differences between countries, building up in Tables 5 and 6 a profile of national institutions which we then use to explain unemployment.

Unemployment benefits

Most EC countries except Italy have benefit systems that are more or less open-ended in duration—unemployed people can draw benefits for at least three years and often indefinitely. By contrast, in the USA and Japan the maximum is half a year and in Norway, Sweden, and Switzerland it is roughly a year. We give summary statistics for 1985 in column (3) of Table 5. In fact, all benefit systems are very complicated (Atkinson and Micklewright 1991). In Annex 3, Table A1, we show exactly which benefits we are counting (i.e. all those paying over $120 a month in 1985).

There is also the question of the replacement ratio. In column (4) we give the replacement ratio over the initial period of unemployment for a single man under 50. This shows gross benefits as a percentage of the most relevant gross wage. As the table shows, replacement ratios are very high in EFTA countries and Denmark and Spain, but the duration is generally limited. In most other countries they are around 50–60 per cent, except for the UK, Australia, and New Zealand, where they are rather lower.

There are two other key dimensions of benefit systems, which are not shown in the table. The first is their coverage—most usefully thought of as the proportion of the unemployed receiving benefit. Table A2 of Annex 3 gives some partial information on this, together with the actual outlays on benefit. Coverage is between a half and three-quarters in most European countries. In the USA it is only a third and in Japan 40 per cent.

The other key issue is the conditions for getting benefit. Such matters are extremely subtle but very important. For example, in Britain virtually no test of work availability was applied in the late 1970s and early 1980s. But from 1986 onwards people receiving unemployment benefit have been interviewed every six months under the Restart Programme and urged to find work. Fewer and fewer reasons are accepted for not taking up available jobs. A strict test of availability for work is also applied to newly unemployed people claiming benefit.

The dramatic fall in British unemployment after 1986 was partially due to these measures, which increased the effective labour supply so that, when demand surged ahead, there was only a limited increase in wage inflation. Corroborating evidence for this interpretation includes the facts that (1) vacancies did not rise despite the fall in unemployment (see Fig. 11); (2) productivity per worker grew at only 1 per cent a year at the peak of the boom; (3) semi-

and un-skilled employment grew strongly; and (4) lower decile earnings fell sharply relative to the mean.

Unfortunately, there are no internationally comparable measures of administrative procedures between countries. But there is a widespread impression in Europe that the 'work test' was applied with progressively less rigour up to the early 1980s, and with rather more rigour since then. And some countries have always been tougher than others. For example, ever since the late 1930s Sweden has consciously adopted what it calls the 'employment principle' as opposed to the 'benefit principle'. This means that unemployed people are expected to look hard for work and, if necessary, to move to get it. In return, they are given major help with job search and in other ways.

Active labour market policy

In fact, countries differ sharply in the amount of 'active' help they give to the unemployed, and not only in the 'passive' help they give through unemployment benefits. Countries vary enormously in what they spend on (*a*) placement and counselling services (plus administration), (*b*) training of adult unemployed, and (*c*) direct job creation and recruitment subsidies. Since the programmes vary with the unemployment situation, the best way to measure a country's commitment to this activity is to measure expenditure per unemployed person (relative to output per worker). As Table 5, column (5) shows, the degree of commitment varies amazingly, with the Swedes doing much more than any other country and Germany doing more than any other EC country. In fact, Sweden goes to the length of guaranteeing every unemployed person a temporary job if he or she has still not found a job when benefits run out (after 14 months).

Table 6 *Collective bargaining in different countries*

	(1) % of workers covered (3 = over 75% 2 = 25–75% 1 = under 25%)	(2) Union coordination (3 = high 2 = middle 1 = low)	(3) Employer coordination (3 = high 2 = middle 1 = low)
Belgium	3	2	2
Denmark	3	3	3
France	3	2	2
Germany	3	2	3
Ireland	3	1	1
Italy	3	2	1
Netherlands	3	2	2
Portugal	3	2	2
Spain	3	2	1
UK	3	1	1
Australia	3	2	1
New Zealand	2	2	1
Canada	2	1	1
USA	1	1	1
Japan	2	2	2
Austria	3	3	3
Finland	3	3	3
Norway	3	3	3
Sweden	3	3	3
Switzerland	2	1	3

Source: cols. (1)–(3): see Annex 4; cols. (4)–(5): *ILO Yearbook of Labor Statistics*; col. (6): Bruno and Sachs (1985: Table 11.7) with minor adjustments; col. (7): OECD, GDP deflator. See Annex 6.

Unions and wage bargaining

As we know, unemployment depends not only on the treatment of the unemployed outsiders, but also on the institutions through which wages are determined, and on how far

(4) Workers involved in strikes p.a. (per 100 workers) (1980s)	(5) Working days lost p.a. (per 100 workers) (1980s)	(6) Wage contract flexibility (index)	(7) Change in inflation 1983–88 (% points)
—	—	4	–3.6
4.8	21.9	6	–3.0
0.8	6.3	3	–6.5
0.7	3.5	4	–1.7
4.5	43.5	2	–7.6
36.3	72.0	4	–8.9
0.5	1.4	5	–0.1
			–12.7
15.0	60.5	1	–5.8
4.6	37.8	2	1.4
11.5	37.5	6	1.1
11.6	47.1	6	1.4
3.3	56.7	2	–1.8
0.5	12.5	1	–0.5
0.4	0.5	4	–0.3
0.3	—	4	–1.9
15.2	50.0	3	–1.6
0.7	10.8	4	–3.5
2.9	18.5	4	–3.8
0.02	—	0	–0.3

these are dominated by insider power. In Europe unions are pervasive in wage-setting, and the percentage of workers unionized rose in most countries up to 1980, since when it has fallen in a few, especially Britain. Union membership is higher in EFTA than in the EC, but in most European countries over three-quarters of workers have wages that are covered by a collective agreement. This is shown in column (1) of Table 6.

But what matters about unions is not only whether they exist, but how centralized they are and thus who is represented in the typical union bargain. In the Nordic countries and Austria the unions operate in a highly centralized way with multi-industry national agreements. In the EC the basic system is for single-industry agreements, which are generally binding on all firms, whether they are unionized or not; however, employers may pay wages above what the industry agreement requires. So it is important whether firm-level strikes over wages are allowed. In Scandinavia, Germany, the Netherlands, and Portugal, they are not.

Within the EC there are big differences between countries in the degree of inter-industry co-ordination that occurs before the industry bargains begin. In Germany, for example, there is a major debate over what 'going rate' makes sense, which runs both in public and in private between the employers' and trade union federations. This leads to a pattern settlement in one industry (in one region) which is then broadly followed elsewhere. Britain is one of the least co-ordinated countries in the EC, with industry-level bargains being of minor importance and little discussion about the going rate. The system in Switzerland is also decentralized but with some employer co-ordination, and with industry-wide peace agreements outlawing firm-level strikes.

Australia and New Zealand have generally had centralized quasi-judicial setting of basic wages, modified by firm-level bargaining about 'over-award' payments. In Japan, wages in large firms are set by synchronized firm-level bargains (preceded by much general discussion), but there is also a large small-firm sector where wages are set by the employer. Finally, in Canada and the USA bargains are at firm level, but the majority of wages in the economy are set at the employer's discretion.

The systems are described in more detail in Annex 4. We need to classify them in a simple way that is yet the most relevant to explaining wage pressure. As we have seen, where union coverage is high, the key issue is whether unions bargain at the national level (thus taking into account the common interests of the workforce in full employment) rather than bargaining as atomistic groups of insiders (thus ignoring the effects of their actions on the general job situation or on the general price level). Of course, even where bargaining is not centralized, if the separate unions agreed on a common wage claim, this would have a similar effect.

But equally, or more, important is the employers' response. If they adopt a common position, then they will certainly not wish to concede real wages high enough to imperil full employment and thus profits. On the other hand, if they bargain one by one they will be more inclined to leapfrog each other, thinking they can achieve some efficiency wage advantage while passing on the cost in an increase of their relative prices. Thus, employer co-ordination could be even more important than union co-ordination.

We therefore construct in Table 6 crude indices of the levels at which unions co-ordinate their wage claims and employers co-ordinate their wage offers: 3 means essentially at national level, 2 at intermediate level and 1 at firm level (i.e. unco-ordinated).

Next, the table records, for interest, two measures of strike activity in the 1980s. Most of the differences are long-standing, reaching back to the Second World War, except for France, where the relative strike record has improved. Strikes are not of course a structural variable, and we shall not use them for explanatory purposes. But it gives some idea of the remarkable differences in industrial relations between countries.

Finally, there is the question of contract structure, which affects the degree of nominal inertia in an economy. If wage contracts are long, then, when nominal demand changes, current wages respond little, and unemployment changes a lot. This effect is reduced if wages are indexed, and it is also reduced if contracts are synchronized rather than overlapping. So we need an index of the extent to which contracts are flexible in the sense of being (*a*) short, (*b*) indexed, and (*c*) synchronized. If we award marks of between 0 and 2 on

each of these points and then add, we have an index of contract flexibility, as shown in column (6).

Explaining cross-section differences

We can now estimate equation (11) as a cross-sectional
equation for the percentage unemployment rate 1983–8 in
each of 20 countries. The results are as follows (with *t*-
statistics in brackets):

Unemployment rate (%)
$$= \quad 0.24(0.1)$$
$$+ \ 0.92(2.9) \text{ benefit duration (years)}$$
$$+ \ 0.17(7.1) \text{ replacement ratio (\%)}$$
$$- \ 0.13(2.3) \text{ active labour market spending (\%)}$$
$$+ \ 2.45(2.4) \text{ coverage of collective bargaining (1–3)}$$
$$- \ 1.42(2.0) \text{ union co-ordination (1–3)}$$
$$- \ 4.28(2.9) \text{ employer co-ordination (1–3)}$$
$$- \ 0.35(2.8) \text{ change in inflation (\% points)}$$
$$\bar{R}^2 = 0.91; \quad \text{s.e.} = 1.41; \quad N = 20$$

Thus, with six institutional variables plus the change in
inflation, we can explain over 90 per cent of the differences
in unemployment between countries.

As one would expect, the duration of benefit is important
(we treated 'indefinite' as four years), and so is the replacement ratio. But it also helps if countries train their unemployed and take active steps to induce or provide work for
them. On the bargaining side, high coverage of collective
bargaining is bad for employment unless it is accompanied
by co-ordinated bargaining. Co-ordination among employers is particularly important. If there is the maximum co-
ordination, as in Scandinavia, a fully covered country can
have lower unemployment than a country with very low
coverage, where efficiency wage considerations may induce
employers to leapfrog.

As it happens, the standardized regression coefficients are

all about one-tenth of the t-statistics quoted above. So the t-statistics indicate well the partial contribution of the different variables to explaining the unemployment differences. These differences are thus explained in roughly equal measure by the treatment of the unemployed, and by the bargaining structures.

Dynamic analysis

But this analysis does not explain why unemployment has changed over time, or why its movement has differed between countries. There are two key points here.

1. Unemployment has moved over time because of supply shocks (changes in γ_0, including now changes in real import prices) and demand shocks (changes in $\Delta^2 m$, the rate of nominal income growth). The extent of these shocks differs between countries.
2. The effect of any given shock depends on the country-specific parameters of the wage and price equations, which in turn depend on the institutional structure of the country.

Real and nominal wage rigidity

We begin with the second of these points. For this purpose we need to modify slightly our initial wage and price equations to allow for the fact that nominal inertia differs between countries. For example, where there are long-term, staggered wage contracts with no indexation, changes in inflation will have a much bigger effect on the mark-up of wages over prices. Thus the wage equation becomes

$$w - p = \gamma_0 - \gamma_1 u - \gamma_2 \Delta^2 p,$$

with a high γ_2 indicating a high level of nominal inertia or nominal rigidity. Similarly, the price equation will be

$$p - w = \beta_0 - \beta_1 u - \beta_2 \Delta^2 p.$$

This gives a Phillips curve

$$\Delta^2 p = -\frac{\beta_1 + \gamma_1}{\beta_2 + \gamma_2}\left(u - \frac{\beta_0 + \gamma_0}{\beta_1 + \gamma_1}\right). \tag{12}$$

Thus, as before, the NAIRU is

$$u^* = \frac{\beta_0 + \gamma_0}{\beta_1 + \gamma_1}.$$

The term $\beta_1 + \gamma_1$ reflects the degree of real wage flexibility in the economy, i.e. the degree to which extra unemployment reduces the gap between the target and feasible real wage. By the same token, the inverse, $1/(\beta_1 + \gamma_1)$, reflects the degree of real wage rigidity (RWR):

$$RWR = \frac{1}{\beta_1 + \gamma_1}.$$

This parameter is very important for two reasons. First, it tells us how the NAIRU responds to a given supply shock, i.e. a given increase in real wage push. For

$$u^* = RWR(\beta_0 + \gamma_0),$$

so that unemployment rises more the greater the degree of real wage rigidity.

Second, real wage rigidity helps to explain how strongly nominal inflation responds to unemployment. For

$$\Delta^2 p = \frac{1}{RWR(\beta_2 + \gamma_2)}(u - u^*),$$

so that the response of inflation is inversely proportional to RWR times the level of nominal inertia ($NI = \beta_2 + \gamma_2$). It is natural to call this latter term ($RWR \cdot NI$) the degree of *nominal* wage rigidity (NWR):

$$NWR = RWR \cdot NI,$$

with

$$\Delta^2 p = - \frac{1}{NWR} (u - u^*).$$

Thus, if a country wants to reduce inflation by 1 percentage point per year, it must experience NWR extra percentage points of unemployment that year. Thus, NWR is often referred to as the 'sacrifice ratio':

$$- \frac{\mathrm{d}u}{\mathrm{d}\Delta^2 p} = NWR = \text{sacrifice ratio.}$$

Countries differ widely in both their real and their nominal wage rigidities. From our estimates of the wage and price equations, we arrive at the numbers shown in Table 7. Real wage rigidity is much higher in most EC countries than in Japan and EFTA, with the USA lying in an intermediate position. However, when it comes to nominal wage rigidity, the USA and Canada rank high, partly because of the prevalence of long-term contracts.

The impact of the oil shocks

The importance of real wage rigidity can be seen at once, if we wish to explain inter-country differences in the impact of the oil shocks. From equation (12) we can see that, for a given wage shock $\Delta\gamma_0$, the change in unemployment over time (Δu) is given by

$$\Delta u = - \frac{\beta_2 + \gamma_2}{\beta_1 + \gamma_1} \Delta(\Delta^2 p) + \frac{1}{\beta_1 + \gamma_1} \Delta\gamma_0.$$

So for the ith country with its own parameter values,

$$\Delta u_i + NWR_i \Delta(\Delta^2 p)_i = RWR_i \Delta\gamma_{0i}. \tag{13}$$

We can use this framework to explain inter-country differences for 19 OECD countries in the change in unemployment between the period 1969–73 and the period 1980–5. For this purpose we focus simply on the effect of the change in relative import prices between 1972 and 1981, so that

$$\Delta\gamma_{0i} = s_{mi}\Delta\log(P_m/P)_i.$$

Using the values of RWR_i and NWR_i from Table 7, the correlation of the left-hand side of equation (13) with the right-hand side of equation (13) is no less than 0.84 (with one dummy variable for Spain).

This analysis gets us only part of the way, however. It falls short in two respects. First, it focuses on only a limited period. We really need a model that explains the year-

Table 7 *Real and nominal wage rigidity*

	Real wage rigidity (RWR)	Nominal wage rigidity (NWR)
Belgium	0.25	0.04
Denmark	0.58	0.08
France	0.23	0.20
Germany	0.63	0.49
Ireland	0.27	0.31
Italy	0.06	0.14
Netherlands	0.25	0.24
Spain	0.52	0.56
UK	0.77	0.70
Australia	1.10	0.10
New Zealand	0.23	0.22
Canada	0.32	1.37
USA	0.25	0.80
Japan	0.06	0.05
Austria	0.11	0.46
Finland	0.29	1.01
Norway	0.08	0.37
Sweden	0.08	0.39
Switzerland	0.13	0.41

Source: Layard *et al.* (1991), Chapter 9, Table 2.

to-year dynamics of unemployment over, say, 30 years, allowing for all kinds of supply and demand shocks. Second, we need to explain the parameter values in terms of the institutional features of each country.

A general dynamic equation

With this in mind, we first obtain the reduced-form equation for unemployment by combining the short-run supply curve (Phillips curve) with the demand curve. The short-run supply curve is (12), expanded to include a term in lagged unemployment, u_{-1}, to allow for an effect of the change in economic activity upon wage and price behaviour (see p. 26). The demand curve is equation (4):

$$\Delta p = \Delta m + \lambda(u - u_{-1}). \qquad (4)$$

After one minor modification,[15] this yields an equation of the form

$$u = bu_{-1} + (1 - b)RWR(a_0 + a_1 z' - NI\Delta^2 m), \qquad (14)$$

where $(\beta_0 + \gamma_0) = a_0 + a_1 z'$, z' being the wage-push factors. Here b is the hysteresis coefficient, which itself increases with real wage rigidity—if unemployment does not reduce wage push, it will not reduce inflation and will therefore persist.

Thus, equation (14) ought to explain the unemployment history of every country, provided we let the coefficients vary according to the factors that should affect them. After suitable experimentation, we conclude that

		Effect
b depends on	benefit duration	+
	co-ordination	–
	labour turnover	–
RWR depends on	benefit duration	+
	co-ordination	–
NI depends on	wage flexibility	–

a_0 varies between countries but a_1 does not. The wage-push variables (z) vary over time and include the replacement ratio, a wage-militancy dummy from 1970 onwards, and $s_m\Delta\log(P_m/P)$—this latter variable taking a coefficient unity.

Thus, using pooled time-series cross-section data for 19 OECD countries (excluding Portugal) for 1956–88, we estimate the following equation (i country, t time):

$$u_{it} = b_i u_{i,t-1} + (1 - b_i)RWR_i(a_{0i} + a_1 z'_{it} - NI_i\Delta^2_{mit})$$

where b_i, RWR_i, and NI_i are themselves functions of the variables mentioned above. The detailed results are as follows

$$b_i = 0.87 - 0.24PL2_i + 0.049BD_i - 0.057EMCD_i + 0.062UNCD_i$$
$$RWR_i = 0.63 + 0.15BD_i - 0.063(UNCD_i + EMCD_i)$$
$$NWR_i = 1.43 - 0.13\ WCF_i$$
$$a_i z'_{it} = s_m\Delta\log(P_m/P)_{it} + 0.082\ WM_{it} + 0.50RR_{it}$$

where $PL2$ is the proportion of employees with job tenure less than two years (an index of labour turnover), BD is benefit duration (Table 5), $EMCD$ is employer coordination (Table 6), $UNCD$ is union coordination (Table 6), WCF is wage contract flexibility (Table 6), $s_m\Delta\log(P_m/P)$ is change in real import prices, WM is a wage militancy dummy, and RR is the replacement ratio. All the coefficients are significant and the results are consistent with the crude cross-sectional results presented earlier in this section. The equation works well for most countries bearing in mind that, aside from the country dummies, there are only twelve parameters. Indeed, for the majority of countries it fits better than an autoregression with time trends *estimated for each country separately*.

Explaining unemployment history

The reduced-form equation above provides us with a splendid basis for discussing in greater detail the events of the 1970s and 1980s. The explanation goes as follows.

(a) Import price shocks. Most countries have been subjected to two major upward shocks to import prices—the first and second oil and commodity price shocks. The size of these shocks was much greater in Europe than in the USA, since the USA produces so much of its own raw materials. However, the more centralized countries suffered less than others because wage-bargainers were more willing to allow the shock to cut living standards.

(b) Demand shocks. By 1980 world inflation had reached a level where electorates in all countries signalled that a change was needed. Most countries restricted the rate of money growth, and most EC countries had severe budget cuts. Thus the growth rate of nominal income fell. Because of inflation inertia, this led not only to a fall in inflation but also to a fall in output and to rising unemployment. The real impact of a given cut in the growth of nominal income was less, the more flexible was the structure of wage contracts, so that for a time unemployment rose as much in the USA (which has inflexible contracts) as in the EC.

(c) Persistence. But in the EC unemployment persisted, while in the USA it fell rapidly after 1982. This was because persistence is much higher where benefits are open-ended in duration. The EFTA countries escaped persistent unemployment partly because unemployment rose little in the first place and partly because persistence there is low, for three main reasons: a limited duration of benefits (in most of the countries), a corporatist approach to wage-setting, and (especially in Sweden) intensive labour market policies for the unemployed.

(d) Other factors. There are also other factors at work which account for some of the deterioration in the unemployment–inflation trade-off. Of these we have been able to identify only the greater militancy of workers after the Paris events of 1968 and rising benefit replacement ratios in many countries at various times up to around 1980.

The main interest is in the policy implications. The clear message is that benefits, labour market policy, and bargaining structure play an important role in affecting the course of unemployment.

12

How Can Unemployment be Reduced?

By bringing together all we have learned, we can now draw significant policy conclusions. Unemployment is not determined by an optimal process of allocation. Though it does perform a vital role in the redirection of labour, its level is subject to a host of distorting influences, tending to make it higher than is economically efficient. The most obvious of these distortions are

1. the benefit system, which is subject to massive problems of moral hazard (unless administered well), and
2. the system of wage determination, where decentralized unions and employers both have incentives to set wages in a way that generates involuntary unemployment, and where bargained wages create a mismatch between the pattern of labour demand and supply.

Both these systems generate negative externalities. While there may be some positive search externalities from unemployment, it is hard to suppose that these are of the same order.

However, the negative distortions do not mean that unemployment is too high in every country. This depends on how much else the country has done to offset them.

Policy-makers have to apply a cost–benefit approach to each possible policy option open to them in their existing circumstances. They inevitably operate in the world of the

second-best and most of the forms of intervention that are
proposed introduce other distortions. Even so, they may
improve the welfare of millions and make an economy
thrive rather than limp.

We shall begin by looking at policies towards the unemployed, including policies on benefits, since the lessons here
are clearest. We shall then look at the issue of bargaining
and incomes policy. Then we shall discuss the role of
employment subsidies.

All these kinds of policies can help a lot. We end by discussing ones that are unlikely to help—profit-sharing,
work-sharing, early retirement, and reduced employment
protection.

Policies for the unemployed (benefits and active manpower policy)

(i) Benefits

The *unconditional* payment of benefits *for an indefinite
period* is clearly a major cause of high European unemployment. This possible effect of the welfare system was never
intended by its founders. For example, the architect of the
British welfare state, Lord Beveridge, proposed in his
Report (1942) that 'unemployment benefit will continue at
the same rate without means test so long as unemployment
lasts, but will normally be subject to a condition of atten-
dance at a work or training centre after a certain period
. . . The normal period of unconditional unemployment
benefit will be six months.' He believed that, after that,
'complete idleness even on an income demoralises'.

Yet somehow this simple truth got overlooked. The
unconditional welfare system worked so well in the boom-
ing 1950s and 1960s that people failed to realize that it
gravely weakened the economy's self-correcting mechanism
in the face of adverse shocks.

The obvious lesson is that unconditional benefits must be

of limited duration. But then, what after they run out? One approach is nothing, as in the USA. This is a harsh route, in which some people end up on the scrap-heap. It also ignores the fact that benefits of even limited duration are subject to 'moral hazard' and liable to encourage an inefficient degree of unemployment. The other approach is active manpower policy.

(ii) Active manpower policy: the Swedish example

The classic example of an active manpower policy is the Swedish system. In the 1960s most foreign economists (including some of us) thought the Swedes had gone over the top. But the wisdom of their approach was proved by the fact that, even after two oil shocks, the Swedish unemployment rate never lingered over 3 per cent; long-term unemployment was never allowed to emerge, and unemployment quite soon came down to under 2 per cent. So it is worth describing the essential features of their system of manpower policy.

Benefits for the unemployed run out after 14 months, but linked to this are labour market policies to make sure that people find productive work. These have four main ingredients.

(a) The placement services (employment exchanges). These go into intensive operation from the moment a person becomes unemployed. Case loads are low—only 35 unemployed people per member of staff, compared with at least five times more in Britain. And the exchanges have excellent information on the labour market both locally and elsewhere, based on the compulsory notification of vacancies.

(b) Retraining. Hard-to-place workers are sent on high-quality training courses—in some cases, as soon as they become unemployed. Thus, economic change is welcomed as an opportunity to provide experienced workers for the

industries of the future. Generally about 1 per cent of the workforce are on courses of this kind.

(c) Recruitment subsidies. If workers have not been placed within six months, employers recruiting them can be offered a 50 per cent wage subsidy lasting six months. The numbers taken up under this scheme peaked at 0.3 per cent of workers in 1984.

(d) Temporary public employment and the right to work. If all these measures fail, the public sector (mainly local authorities) acts as the employer of last resort. It provides work for up to six months, mostly in construction or the caring services. Provision is highly counter-cyclical, covering some 2 per cent of the workforce at the peak and under 0.5 per cent by 1988. Anyone whose benefit entitlement has run out is entitled to such work by law.

Such policies are expensive, and the Swedes spend nearly 1 per cent of national income on them. But, by keeping down unemployment, the programmes reduce unemployment benefits, which in the EC cost 1.5 per cent of GNP compared with 0.7 per cent in Sweden. In the long term the Swedish programmes may be largely self-financing to the Exchequer. In terms of social cost–benefit analysis, they almost certainly pass the test.[16]

By any criteria, the Swedish labour market has performed extremely well (during the 1980s). The employment–population ratio, already the highest in the world, has gone on rising, while it has fallen in all the main EC countries. However, in the last year or so, the system has been placed under extreme pressure. As a consequence of various adverse demand shocks and problems with the banking system, unemployment in Sweden has risen to unprecedented heights (over 5 per cent). Whether or not the Swedish system can survive intact under this pressure remains to be seen.

(iii) Policy towards the long-term unemployed

The lessons here are particularly obvious for the countries of Eastern Europe which have started from a position of zero unemployment. But, for a country with high unemployment, there is also the question of how to get from here to there. In high-unemployment countries around half the unemployed have been out of work for over a year. For such workers the chances of finding a job are very much less than for the short-term unemployed. And, for the same reason, long-term unemployment is doing much less to restrain inflation than short-term unemployment.

For these reasons, active help to the unemployed should be concentrated on the prevention of long-term unemployment. If we remove from unemployment a newly unemployed person, we are removing someone who on average would have left unemployment fairly soon anyway. If we remove a person at risk of long-term unemployment, we are removing someone who might otherwise continue much longer in unemployment. So the external benefit to the taxpayer from removing the second type of person is much greater than that from removing the first. Unless the costs are disproportionately greater, therefore, help should be concentrated on those at risk of long-term unemployment.

(iv) Displacement, substitution, and deadweight

But manpower policies are often criticized on two grounds. First, though they provide jobs for those helped, they may reduce employment for others (by 'displacing' labour in other firms, or 'substituting' for other workers in the same firm). This argument is often based on the notion that there is a limited demand for labour (arising from limited aggregate demand for products). If so, the argument is almost totally misconceived. For the aim of manpower policy is to improve the supply side of the economy, on the assumption that this is the main limiting factor, not aggregate demand.

But there will almost certainly be some substitution and displacement for supply-side reasons. For example, if long-term unemployment is greatly reduced, there may need to be some small increase in short-term unemployment in order to restrain wage pressure. In principle, the magnitude of the total effect of a policy can be determined by finding out how it affects not only the outflow rate from unemployment but also the inflow rate (the unemployment rate being determined by the ratio of the inflow rate to the outflow rate).[17]

The second charge against manpower policy is that it often pays money for things that would otherwise have happened ('deadweight'); for example, an employer is paid for hiring someone he would have hired anyway. Transfer expenditures of this kind are undesirable if they then have to be paid for by distorting taxes. But such elements are probably a smallish issue in the overall social cost-benefit calculus of most active labour market policies.

(v) Pin-point targeting

The policies we have discussed have the major merit of being targeted directly at the problem in hand. For example, general regional aid is often advocated because there are more unemployed in one region than another. But much of it fails to relieve unemployment. In contrast, the policies we have been discussing aim directly at unemployment. They are thus highly regional, but they are regional *as a consequence* of dealing with unemployment, rather than in order imperfectly to do so. Likewise, these policies deal directly with skills mismatch where it is identified, rather than by some more general intervention.

Policies on mismatch (employment subsidies and training)

This does however raise the issue of whether there is a case for more general action to combat the mismatch across regions and across skills. Suppose there are two markets (say North and South), with higher unemployment in the North. One could approach this problem by increasing labour demand in the North or by reducing labour supply there (by out-migration). But it does not make sense to attempt both; for subsidies to employment in the North will be paid for by higher taxes in the South. This policy is bound to discourage migration.

So which policy should be attempted? If better returns to migration do little to encourage migration, then (ignoring externalities) the correct policy is to subsidize employment in the high-unemployment area. But suppose migration is very responsive, with all workers indifferent between regions at the prevailing rates of wages and employment. Then, even though there is job rationing, the classic principles of public finance apply: in the absence of externalities, there should be uniform taxation.

However, there are externalities. Migration into low-unemployment areas creates a demand for extra infrastructure, publicly financed. It may also damage the losing region. This argument, together with unresponsive migration behaviour, provides the foundation of the case for regional policy. But one must stress that other distortions that reduce migration, such as housing policy, do need urgent reform.

With skill formation, the case is somewhat different. Training suffers from the standard externality problem— that trainers are not able to trap the full social return, either because of 'poaching' or because of the tax wedge. Even though the supply response is again quite weak, this constitutes a case for favourable fiscal treatment for education and training.

As we have already said, direct policies affecting the unemployed should be judged by different criteria from those affecting the overall balance of supply and demand. This is because of the pin-point targeting which gives them their extra leverage.

The reform of wage bargaining, and incomes policy

(i) Bargaining systems

We turn now to the other key issue: the reform of wage bargaining. Here we have discovered two main points. First, other things equal, unemployment is lower the lower is union coverage and the lower is union power in each bargain. This suggests the merits of limiting the power of individual unions. But, second, for a given union coverage and union power, unemployment is lower when employers co-ordinate their wage offers at an industry or national level, and likewise when unions co-ordinate their wage claims.

So there seem to be two forms of organization that work well. One (as in the USA) has low union coverage—and preferably low union power. The other (as in Scandinavia and Austria, and to a lesser extent Germany) has high union coverage—with low union power again at the decentralized level, but with strong national unions dealing on equal terms with employers. The choice between these systems is clearly political and depends also on the size of country. But economic arguments are also relevant.

The issue is whether institutions can be created which overcome the externalities involved in decentralized wage-setting (whether by firms and/or unions). The ideal is that a consensus develops about an appropriate 'going rate' for nominal wages, which is then implemented without requiring unemployment to eliminate the wage–price and wage–wage spirals. In this context there is a role for

1. an informed national debate about what rate makes sense;
2. reports by respected bodies such as councils of economic advisers and research institutes;
3. national talks between employers and unions.

If the climate of opinion is responsible, a kind of implicit contract may emerge, as often happens in Germany and

Japan, in which other bargainers follow a pattern settlement unless they face exceptional circumstances. Everyone recognizes the need for increasing flexibility in remuneration packages. But equally, it is important that most agreements stick within an accepted range of total remuneration and do not initiate a game of competitive leapfrogging.

However, this does presuppose a fairly high degree of social discipline. If this is not forthcoming, governments naturally consider direct intervention.

(ii) Conventional incomes policies

We then need to consider the case for some form of government wage controls, such as a maximum permitted percentage rate of growth of earnings. Incomes policies of this kind have been tried at many times and places.

To control inflation, the Roman Emperor Diocletian issued a wage decree in AD 301 and those who breached it were sentenced to death. The policy was abandoned as a failure after 13 years.

In AD 1971 the US President Nixon introduced a three-month wage–price freeze, followed by two years of less rigid controls. The policy clearly restrained inflationary pressure while it lasted, but proved unsustainable under the pressure of shortages of labour and goods (Blinder 1979).

In Britain there was a statutory incomes policy in 1972–4 and a voluntary one (initially agreed with the Trades Union Congress) in 1975–9. Both of these were abandoned, mainly because of union opposition. However, the second of the policies was at first remarkably successful, and helped to reduce inflation from 27 to 12 per cent in two years with no increase in unemployment. After the policy was abandoned inflation rose again. Some people said this was due to a 'catching-up effect'. But the best econometric evidence does not support the view that in Britain reductions of inflation achieved during incomes policies are automatically undone once the policies end (Wadhwani 1985).

In France an incomes policy was introduced in 1982 and inflation fell over four years from 12 to 3 per cent. The wage norms had statutory force in the public sector, and the employers' federation broadly followed the same norms.

Similarly, Belgium and Italy have, since 1982, had laws prescribing the maximum degree of wage indexation in between major renegotiations, which again implies a form of wage norm. Inflation has fallen.

Australia has a long-standing system of quasi-judicial determination of basic wage rates, above which 'over-award' payments can be negotiated. However, since 1983 the national government, in 'accord' with the union movement, has set the basic norm within which the system operates.

There are two main problems with fully centralized governmental incomes policies. First, they infringe the principle of free bargaining between workers and employers. Thus, many individual groups have a strong incentive to breach the norm. This is also the case, of course, where a norm has been bargained centrally between confederations of employers and unions. But individual groups are more inclined to accept a deal to which they are at least an indirect party. For this reason, governmental incomes policies that have the support of the confederations of employers and unions are themselves more likely to last than those that are imposed. But history suggests that nearly all such policies are eventually breached. A permanent centralized incomes policy is probably infeasible.

The second problem is that a centralized incomes policy is inherently inflexible. It is bound to impose rigidity on the structure of relative wages. But the reallocation of labour may be much easier if relative wages rise where labour is scarce and vice versa. Without this, structural unemployment is likely to become worse, unless major efforts are made, as in Sweden, to promote movement of labour between industries and regions. Incomes policies sometimes try to incorporate committee mechanisms for adjusting relativities, but these cannot work as effectively as the market.

The result is that incomes policies of this kind have

always been short-lived. This does not mean they have always been useless. Indeed, a temporary incomes policy is a much better way to disinflate than having a period of high unemployment. And if unemployment is above the long-run NAIRU and there is hysteresis, a temporary incomes policy is an excellent way of helping unemployment to return to the NAIRU more quickly.

(iii) Tax-based incomes policies

One would, however, like to achieve a permanent reduction in the NAIRU itself. If this is to be through incomes policy, it must be through some mechanism other than direct controls. This leads to the proposal for tax-based incomes policy. Under this there is a norm for the growth of nominal wages, but employers are free to pay more than the norm at the cost of a substantial financial penalty. Thus, if employers need to break the norm in order to recruit labour or avoid a strike, they will do so. But all bargainers will be subject to strong disincentives to excessive settlements. Let us see more clearly how this would work.

If the free market generates excessive wage pressure, the obvious solution is to tax excessive wages. This is generally the most efficient way to deal with market failure, unless direct controls have some particular advantage. One approach is through a tax on excessive wage growth; another is through a progressive tax on wage levels. For the sake of clarity, we shall discuss them in reverse order, starting with a tax on the *level* of wages.

Suppose that the tax is paid by firms. If a firm pays its workers a gross real wage W_i, it also has to pay the Exchequer a net tax per worker of $tW_i - S$, where t is the tax rate and S a positive per worker subsidy. Hence the firm faces an *ex ante* schedule of labour cost per worker (C_i) equal to

$$C_i = W_i(1 + t) - S.$$

We assume that the scheme is self-financing, so that *ex post* in the representative firm $C_i = W_i$.

How does this reduce wage pressure and thus unemployment? The basic mechanism is that, when workers gain an extra £1 of wages, it costs the firm an extra £$(1 + t)$. Thus, the firm is more willing to resist any claim, while the workers may be more anxious about making the claim because of its greater employment effect. As on p. 39, the bargained wage W_i is that which maximizes $\beta\log(W_i - A)S_i + \log\Pi_i$. Differentiating this expression with respect to W_i, the firm chooses the wage so that

$$\frac{\beta}{W_i - A} + \frac{\beta}{S_i}\frac{\partial S_i}{\partial C_i}\frac{\partial C_i}{\partial W_i} - \frac{N_i}{\Pi_i}\frac{\partial C_i}{\partial W_i} = 0,$$

where by the envelope theorem a unit rise in labour cost (C_i) reduces profit by N_i so that $\partial\Pi_i/\partial C_i = -N_i$.

Since the tax sets $\partial C_i/\partial W_i = 1 + t$, and *ex post* it is self-financing with $C_i = W_i$, the mark-up of the wage over outside opportunities (Equation 8) is now given by

$$\frac{W_i - A}{W_i} = \frac{1 - \alpha\kappa}{(1 + t)(\epsilon_{SN} + \alpha\kappa/\beta)}.$$

The higher the tax rate, the less will wages tend to leapfrog each other. Thus unemployment will be lower. To be precise, since $W_i = W = W^e$, equation (8') now becomes

$$u^* = \frac{1 - \alpha\kappa}{(1 + t)(\epsilon_{SN} + \alpha\kappa/\beta)\varphi(1 - B/W)},$$

so that unemployment falls as the tax rate rises. A similar result holds in the case of efficiency wages.

Needless to say, it makes no difference whether the tax is levied on firms or workers.[18] But it must be progressive so that, when wages rise, labour cost rises faster than wages do; i.e., a part of wage cost must be tax-exempt, through a positive S. A proportional tax at rate t whose proceeds were given to the Martians would have no effect.

Of course, any tax introduces some distortions, even while it offsets others. A tax on weekly earnings could have severe effects on work incentives, so the tax should be levied on hourly earnings to make it as near an ideal tax as possible.

An alternative, and more understandable, policy is to tax the *growth* in hourly earnings. The upshot again is lower wage pressure and lower unemployment. But the tax bites less hard, because raising wages this year rather than next costs you taxes this year but saves you taxes next year. Thus, to achieve a given reduction in wage pressure, the tax rate has to be $1/(r - n)$ times what is needed with a wage level tax, where r is the real discount rate and n the permitted (tax-free) growth rate of real wages.

According to many people, a tax-based incomes policy is very difficult to administer. This is not true. Provided it is part of the law of the land and the definition of earnings is as for the income tax (or the social security tax), it can be readily collected from firms at the same time as they pay the withholding income tax (or the social security tax). There are, as with any tax, some obvious ways of trying to dodge the tax. Most of these can be dealt with. Even so, any tax has some distorting effects and so does TIP. But on balance we believe that, if the political will were there to implement it, in most countries it would not only decrease unemployment but would raise social welfare.

We should stress that the aim of all incomes policies is not to reduce real take-home pay but only to reduce wage pressure and thus the NAIRU. Indeed, since higher output yields higher tax returns, it will normally be possible to cut tax rates when employment increases.

Marginal employment subsidies

Incomes policy works by reducing the target real wage at given unemployment. An alternative way to reduce unemployment is to raise the feasible real wage in a way that does not lead to equal changes in the target real wage. A good way to do this is by a marginal employment subsidy.

Suppose that we subsidize at a rate s all employment above some fixed proportion of last period's employment. If the scheme is self-financing, it can be paid for by a tax on the rest of last year's employment. If the firm is monopolistically competitive, it sets prices as a mark-up on marginal cost. Thus the price equation becomes

$$p - (w^e - s) = \beta_0 - \beta_1 u.$$

The feasible real wage is increased and unemployment falls. This is an attractive way of reducing inflationary pressure.

Clearly, we do not want this process to reduce post-tax profits, but post-tax profits can be restored by reductions in the profit tax financed by proportional taxes on workers. The latter, as we have seen, would not affect unemployment.

Another way to reduce the profit mark-up is by increased product-market competition (e.g. via the 1992 programme in Europe). Under wage-bargaining (though not efficiency wages), this will reduce unemployment.

We turn now to policies that are much less likely to have this effect.

Profit-sharing

There has been much recent excitement over profit-sharing, generated by the work of Martin Weitzman. Social reformers have, of course, advocated profit-sharing for many years as a way of improving productivity—and there is good evidence to support their case. But the extra productivity would not of itself increase employment. That would require some additional mechanism.

In his original book, Weitzman (1984) proposed such a mechanism in the context of a labour market which in equilibrium is market-clearing. He argued that under the wage system firms equate the real wage to the marginal revenue product of labour. In the short period the real wage is fixed, so that any fall in marginal revenue product will reduce employment. Under profit-sharing, by contrast, competition for labour ensures that in general equilibrium the marginal revenue product equals the total remuneration of labour (i.e. the base wage plus the profit share). Hence the marginal revenue product exceeds the base wage. But, once the base wage has been set, *ex post* firms seek to employ labour up to where the marginal revenue product *equals* the base wage. So there is permanent excess demand for labour. A fall in labour demand (marginal revenue product) will not cause a fall in employment—merely in profits. Weitzman claimed that this explained the Japanese miracle.

But there are problems with the theory and with the Japanese evidence. The theory assumed that, after the package of base wage and profit share had been determined, workers would stand idly by while the firm tried to employ people, thus eroding the profit share of the existing workers. It seems unlikely that workers would react in this way, rather than trying to bargain also about employment. Second, the theory assumed long-run market-clearing in the labour market. In many countries this may not be the right model, and it is easy to show that in both an efficiency wage model and our bargaining model profit-sharing would have no effect on the NAIRU.

So what about Japan? Why exactly is unemployment in Japan so low and so stable? It is not because of any of the mechanisms Weitzman describes, as the following facts make clear.

1. Output is not stable. It fluctuates (about its trend) more than in most countries. It responds to monetary shocks exactly as elsewhere.
2. Nominal prices are affected by cost factors and not simply by demand.
3. Excess demand for labour, as reported by firms, is rather lower than in other countries.
4. It does not appear that employment is determined in the short run by base wages.

Having said all this, the basic fact remains that employment in Japan *is* stable compared with elsewhere. What happens is roughly as follows. Only 40 per cent of Japanese workers are in the organized sector (where bonuses are paid); another 30 per cent are employees in the small-firm sector, and 30 per cent are family workers. When output fluctuates, employment in the formal sector fluctuates quite a lot. But employment in small firms varies much less. This is quite simply because the flexibility of pay per worker is so high in the market-clearing small-firm sector, while it is much less high in large bonus-paying firms. Thus, Japan's stable employment record is due mainly to the wage flexibility in the small-firm sector.

This flexibility has the result that in Japan the total labour input (hours × employment, HN) fluctuates less than in other countries. On top of this, the Japanese value their human capital highly, so they use hours per worker (H) as a shock-absorber more than most other countries, further dampening fluctuations in employment (N). In addition, the labour force (L) shrinks in recession, as 'secondary' female workers shrink back home. This makes unemployment ($L - N$) even more stable than employment (compared with other countries).

So what does the Japanese evidence tell us about profit-sharing? Since the intermediate predictions of Weitzman's theory are not borne out, one can say either that his theory is wrong or that Japan is not a case of profit-sharing. There is a lot to the latter view. While some 25 per cent of remuneration is in bonuses, much of this is indeed a fixed element. Thus, we must probably conclude that Japan provides little evidence either for or against profit-sharing.

Even so, we would support profit-sharing as a device to improve productivity and industrial relations. As a device to reduce unemployment, it is no straightforward panacea.

Early retirement and work-sharing

Two policies that are very popular would be clearly counter-productive. The first is the policy of reducing the labour force by early retirement. As we have shown, it is the unemployment rate that equilibrates the labour market. If the size of the labour force is reduced, the equilibrium unemployment rate is unaffected. Employment has to fall to eliminate the wage pressure that would otherwise emerge, as the supply of labour becomes more scarce relative to the demand. Thus, early retirement does not make jobs available for people who would otherwise be unemployed: it just reduces employment.

This is what reasonable theory says, and it is confirmed by the evidence. In time-series regressions wage behaviour is affected not only (positively) by employment but also (negatively) by the size of the labour force—and the absolute elasticities are of roughly equal size. Moreover, if one compares countries, it is striking that early retirement has expanded most in countries with the greatest increase in unemployment. In Japan, where retirement behaviour is unchanged, unemployment has not risen at all. This suggests that early retirement is not an effective means of reducing unemployment. It is an excellent way of making a country poorer.

The other policy with the same effect is work-sharing. The idea here is to redistribute the available work to more people. But once again, the available work is not a given— that is the 'lump-of-output fallacy'. The equilibrium unemployment rate is independent of hours of work. Thus, if hours are reduced and employment rises for a while, wage pressure will soon increase and the amount of work available will have to be reduced. Employment will revert to its former level.

We can understand why this happens by taking our wage-setting models, inserting hours, and making W_i represent the hourly wage. The conclusion from theory is confirmed by time-series regressions, which show that hours do not affect the relation between wage pressure and the unemployment rate. Again, the countries that have reduced hours most have been those where unemployment has grown most. In Japan and the USA, with fairly steady unemployment, hours have fallen little. Thus, cuts in hours provide a poor antidote to unemployment. But they certainly provide a lower standard of living.

Employment protection legislation

Another policy of importance relates to the laws of employment protection. In most European countries the law requires that, when a worker is laid off, he be given advance notice, severance pay (redundancy payments), and a satisfactory reason (as opposed to 'unfair dismissal').

Laws of this kind must reduce the rate of flow into unemployment (S/N), and this effect tends to reduce unemployment. But such laws also discourage hiring, since firms are less willing to hire workers whom they cannot later dismiss without incurring costs. Thus, the outflow rate from unemployment (H/U) is also reduced. In equilibrium the outflow from unemployment has to equal the inflow ($H = S$). Thus

$$\frac{U}{N} = \frac{S/N}{H/U}.$$

Unemployment is reduced by employment protection only if the inflow rate (S/N) falls more than the outflow rate (H/U). Studies on this matter yield ambiguous results.

On balance, employment protection laws are probably bad for employment, since they strengthen insider power and encourage the payment of efficiency wages to motivate workers who cannot be threatened with dismissal. But there are equity arguments in their favour, and the evidence on adverse employment effects is not strong enough to warrant a total abandonment of the practice.

Demand management

On the supply side, we have seen that there exist policies which would really help (policies towards the unemployed, towards wage determination, and marginal employment subsidies)—and some others which would probably not. What about the demand side?

This is not mainly a book about the demand side of the economy, or about 'stabilization policy'. We would make only two comments.

First, when hysteresis is strong, it is very important to avoid big rises in unemployment. If inflation is too high, it is better to eliminate it by small amounts of extra unemployment over a longish time period than by anything approaching 'cold turkey' (see Annex 5). Had this been understood in 1980, some of the disaster of European unemployment could have been avoided.

Second, once inflation is at an acceptable level, it is normally desirable to avoid disturbances to nominal demand, by holding the growth of nominal demand stable. But should inflationary supply shocks happen, the case for some accommodation through faster nominal demand growth is stronger the higher the degree of hysteresis. Stabilization policy should be highly sensitive to the supply mechanisms of the economy.

13

Summary

We began with a set of ten questions, which we have taken some time to answer. If we had been quicker, we might simply have said:

1. Unemployment is in equilibrium when it is high enough to eliminate the leapfrogging of wages over each other and to make the planned mark-up of wages over prices (the target real wage) consistent with the planned mark-up of prices over wages (the feasible real wage).

2. There is, however, 'nominal inertia' in price- and wage-setting so that the system can easily depart from equilibrium as a result of shocks. Moreover, once unemployment is away from the long-run NAIRU, it takes some time to return. If recent unemployment is high, inflation falls only if unemployment is above the *short-run* NAIRU.

3. In the steady state, lower unemployment may require lower real labour costs but not necessarily lower real take-home pay.

4. Equilibrium unemployment is not market-clearing. Firms may find it profitable to pay wages above market-clearing levels in order to motivate workers. Unions may also keep wages up, even when there is excess supply of labour.

5. Unemployment is raised by adverse demand shocks and adverse supply shocks (such as rises in relative import prices). But in the very long term unemployment is independent of import prices, taxes, and productivity.

6. But unemployment is also affected by the search behaviour of the unemployed, and is higher when the unemployed search less (whether because of unemployment

benefits or because of the demoralization arising from long-term unemployment).

7. For some people there is also a secondary sector of the labour market, where wages clear the market and jobs are available. If workers are not taking these jobs, it is because the jobs are too unpleasant or ill-paid relative to the quality of life while unemployed.

8. Unemployment rates are much above average for less skilled workers, young people, and people in disadvantaged regions. These disparities tend to raise the overall unemployment rate.

9. The different experience of different countries depends on the way they treat unemployed people (benefits and active manpower policy) and their wage-bargaining systems—together with the shocks they have been subjected to.

10. There is plenty we can do to reduce unemployment. It is far from natural, and not beyond our power to control.

Notes

1. This is related to the finding that, when unemployment is regressed on lagged unemployment, the coefficient on the latter is close to unity. For Britain, annual data for 1900–89 give

$$u_t = 0.0041 + 0.934\, u_{t-1}$$
$$(0.039)$$

   and for the USA they give

$$u_t = 0.0080 + 0.877\, u_{t-1}$$
$$(0.051)$$

   (s.e. in brackets). However, this kind of exercise assumes an unvarying stochastic 'unemployment process', which we do not.
2. The equation corresponding to Fig. 5(c) is

$$\Delta^2 p = 2.55 - 0.79u + 0.15t$$
$$(4.0) \quad (2.8)$$

   where $\Delta^2 p$ is the annual change in inflation (% points), u is the unemployment rate (%) and t = date—1990. Clearly, a time trend is a very inadequate way to model supply shocks, and most of this book is concerned with seeking better ways to do so.
3. The long-run equilibrium rate of unemployment is also often called the 'natural' rate (Friedman 1968). We avoid this usage which smacks of inevitability.
4. A more accurate term would be non-increasing inflation rate of unemployment, but the common usage is NAIRU.
5. If wage and price behaviour depend on changes in unemployment, equations (1), (2) become

$$p - w^e = \beta_0 - \beta_1 u - \beta_{11}\,(u - u_{-1}); \quad w - p^e = \gamma_0 - \gamma_1 u - \gamma_{11}\,(u - u_{-1})$$

   If we add these together, and assume that wage and price surprises are similar and equal to the change in inflation, $\Delta^2 p$, we have

$$\Delta^2 p = -\theta_1(u - u^*) - \theta_{11}\,(u - u_{-1})$$

   where $u^* = (\beta_0 + \alpha_0)/(\beta_1 + \gamma_1)$, $\theta_1 = \frac{1}{2}(\beta_1 + \gamma_1)$, $\theta_{11} = \frac{1}{2}(\beta_{11} + \gamma_{11})$.
6. If u_{-2} also affects Δp, the short-run NAIRU may lie above u_{-1} if unemployment has risen rapidly.

7. During disinflation the observations are in the corresponding area to the left.

8. Solving (1') and (2') for $w - p$ shows that

$$w - p - (w^* - p^*) = \frac{\gamma_1(w - w^e) - \beta_1(p - p^e)}{\beta_1 + \gamma_1}.$$

Suppose that in the first period after a shock $w^e = w^*$ and $p^e = p^*$; then

$$\frac{w - w^*}{p - p^*} = \frac{\gamma_1}{\beta_1}$$

so that

$$\Delta \log w > \Delta \log p \text{ if } \gamma_1 > \beta_1.$$

9. Millward and Stevens (1986), Table 8.8. The redundancy figures are based on Layard *et al.* (1991), Chapter 2, Table 5.

10. The consumer price is

$$P_c = P^{1 - s_m} P_m{}^{s_m} = P(P_m/P)^{s_m}$$

11. Annex 1.2 gives a model of the world economy, endogenizing commodity prices.

12. As we explain in the next section, hires, H, are increasing in both the number of vacancies, V, and the number of effective job-seekers, cU. Thus we have a relation of the form $H = h(V, cU)$. Under reasonable assumptions, h has constant returns and hence $1 = h$ $[V/H, cU/H]$. Consequently, the chance of filling a vacancy, V/H, is uniquely related to the chance of getting a job, H/cU.

13. It is also probably the case that changes in primary employment cause almost equal changes in unemployment (though not of course quite one for one).

$$-\frac{dU}{dN_1} = 1 - \frac{N_2}{N_2 + U} \frac{1}{1 + \eta^S/\eta^D}$$

where η^S is the elasticity of S_2 and η^D is the elasticity of D_2, both evaluated at the point of intersection. If η^S is high, this is close to one.

14. Set $u_i/u = x$. Expanding x around $\bar{x}$ gives

$$\log x \simeq \log \bar{x} + \frac{1}{\bar{x}}(x - \bar{x}) - \frac{1}{2}\frac{1}{\bar{x}^2}(x - \bar{x})^2.$$

Since $\bar{x} = 1$ and $\Sigma \alpha_i = 1$,

$$\Sigma \alpha_i \log x_i \simeq 0 + 0 - \frac{1}{2}\Sigma \alpha_i (x_i - \bar{x})_2.$$

Thus, provided $(\alpha_i - L_i/L)$ is independent of u_i/u,

$$\Sigma \alpha_i \log u_i / u \simeq -\frac{1}{2}\Sigma \frac{L_i}{L}\left(\frac{u_i}{u}-1\right)^2.$$

15. We replace Δp_{-1} by Δm_{-1}. This yields the correct result under rational expectations (see Layard *et al.* 1991, Chapter 9).
16. Control group studies of the effect of the programmes on individuals do not show a totally clear pattern of results (Björklund 1990). But two caveats are important. First, there have been no experiments using random assignment. But, second, the programmes may help to create a pro-work ethic and may have important externalities that can only be judged by looking at the overall performance of the system compared with other countries, and not by comparing the experience of one Swede with another. For a full description of the Swedish system see Layard and Philpott (1991).
17. The inflow rate is inflow/N and the outflow rate is outflow/U. In flow equilibrium, inflow = outflow and the ratio becomes U/N.
18. To analyse a tax on workers, make take-home pay (W_i) equal to $C_i(1 - t) + S$ and differentiate the objective function with respect to W_i.

Annexes

1. The 'intertemporal substitution' theory of fluctuations

IF the labour market always cleared, we should have to be able to explain why more people are willing to work in booms than in slumps. The first (and most widely used) explanation is based on the notion of intertemporal substitution (Lucas and Rapping 1969). According to this, workers work harder in years when the perceived real wage or the perceived real interest rate is unusually high (making it worth working more now to consume more later).

The basic empirical problem with this is that neither the real wage nor the real interest rate is strongly pro-cyclical. And many would argue that the real wage is not far from a random walk with drift, so that the best forecast of the future real wage is the current real wage plus trend (Altonji and Ashenfelter 1980). In any case, if the theory is to fit the facts, the intertemporal elasticities of substitution must be large, which they do not appear to be (see e.g. Altonji 1986). In addition, consumption and employment move strongly together over the cycle, which, for consistency with the theory, requires non-intertemporally additive utility functions (since otherwise consumption in each period is complementary with leisure).

Thus, not surprisingly, tests of the model have been generally unfavourable (Altonji 1982; Ashenfelter 1984; Abowd and Card 1989). Perhaps the most decisive rejection comes from the work of Ham (1986), who used panel data. His key finding was that almost all the unemployed people worked far less than the estimated model predicted. Thus, the model failed to perform its main job of explaining unemployment. On aggregate time-series data on labour supply and consumption, Mankiw et al. (1985) also comprehensively reject the model.

One obvious shortcoming of the model is that it explains hours of work rather than employment versus non-employment. In an effort to rectify this, Hansen (1985) and Rogerson (1988) have introduced fixed costs of work which of course include the loss of unemployment benefit. Because of this, there is a reservation wage below which the hours an individual is willing to work fall discontinuously to zero. This is an important and extremely well-known fact about life. It certainly explains why, where there is a market-clearing secondary sector, not everyone is willing to work in it (see below). But, as noted above, the important fluctuations in employment originate in the primary sector, where wages exceed fixed costs. Consequently this idea helps little in understanding unemployment movements.

In any case, intertemporal substitution could only explain temporary fluctuations in employment. The notion that it could explain the persistent high unemployment of the 1930s or 1980s cannot even be contemplated.

2. A model of the OECD economy with endogenous commodity prices

From the point of view of any one (small) country, a rise in the relative price of oil and other commodities is an exogenous supply shock. But at the level of the OECD as a whole, it is endogenous. The relative OECD terms of trade P/P_m vary negatively with the level of OECD activity $(1 - u)$. This effect is quite marked in the short run, but weak (or zero) in the long run. The short-run and long-run relationships are shown in Fig. A1 as *SRCP* and *LRCP* respectively (for short-run and long-run commodity prices).

The relative price of commodities in turn affects the level of OECD activity consistent in the short run with stable inflation. For, from equation (3'''') in Chapter 7, with $\theta_{11} = 0$ and Δwedge $= s_m \Delta \log P_m/p$,

$$1 - u = 1 - u^* + \frac{1}{\theta_1} s_m \Delta \log P/P_m + \frac{1}{\theta_1} \Delta^2 p.$$

The short-run relationship is shown as *SRS* (for short-run supply) and the long-run (vertical) relationship as *LRS* (long-run supply). In the short run all points to the right of *SRS* are points of increasing inflation.

In 1973–4 world economic activity became so high that commodity prices rose to a level (such as A) where inflation was bound to rise. In 1979, the fall of the Shah of Iran induced a downwards shift in the supply of commodities. With OECD output slow to respond downwards, increasing inflation was again inevitable (as at B).

For an empirical analysis along these lines, see Cristini (1989).

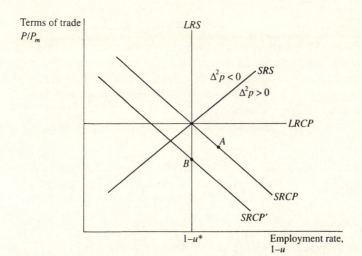

Fig. A1. *Relation of commodity prices and the OECD NAIRU.*
SCRP, LRCP refer to the determination of commodity prices;
SRS, LRS refer to the constant-inflation supply price of output.

3. Unemployment benefit systems in OECD countries

See table over
Notes: Col. (1): Duration of eligibility to some form of benefit
paying over $120 a month. Equals sum of cols. (3) + (5) + (7),
except for Sweden (see below). In Japan, 90–300 days depending on eligibility: in 1987 weighted average was 166 days
(Koyo Hoken Jigyo Tokei, *Unemployment Benefit Statistics*,
1987). Col. (2): Gross benefits for a single person under 50 as
% of the most relevant wage, normally gross wage. All systems
pay the indicated proportion of former earnings (up to a ceiling generally exceeding average earnings), except where noted
below. (In Germany data relate to benefits relative to net
wage.) Gross benefits are normally taxable. *Austria*: The rate
is 30–60%, depending on earnings. *Britain*: £28.45 a week. But
those receiving insurance were also eligible for means-tested
assistance (Supplementary Benefit) which paid their rent.
Taking the relevant average rent as £10 a week gives a total
equal to 36% of lower quartile weekly earnings of full-time
workers (£107). *Finland*: FM70 a day indefinitely plus earn-
ings-related additions for 21 mos. *France*: Fr40 a day plus 42%
of earnings (up to 5 times average earnings). *Italy*: We do not

Table A1 *Unemployment benefit: duration and replacement ratios: single person, 1985*

	Maximum duration of benefits (yrs.) (1)	Unemployment insurance	
		Replacement ratio (%) (2)	Duration (yrs.) (3)
Belgium	Indef.	60	1
Denmark	2.5	90	2.5
France	3.75	57	3.75
Germany	Indef.	63	1
Ireland	Indef.	50	1.25
Italy	0.5	2	0.5
Netherlands	Indef.	70	2.5
Spain	3.5	80	0.5
UK	Indef.	36	1.0
Australia	Indef.	—	—
New Zealand	Indef.	—	—
Canada	0.5	60	0.5
USA	0.5	50	0.5
Japan	0.5	60	0.5
Austria	Indef.	60	0.6
Finland	Indef.	75	Indef.
Norway	1.5	65	1.5
Sweden	1.2	80	1.2
Switzerland	1.0	70	1.0

include the Cassa Integrazione Guadagni (Wage Supplementation Fund). This pays 80% of the salary of temporary layoffs, but only about 1% of the workforce was on the scheme in 1988. Basic unemployment benefit was L800 per day. *Ireland*: I£39.50 p.w. plus 20–40% of earnings over I£36 depending on wage (up to 1.2 × ave. earnings). Maximum is 85% of net earnings. *Japan*: 60–80% of former earnings (excl. bonus). In 1988 the weighted average was 63%. Average benefit paid per

Table A1—*continued*

	Supplementary insurance		Means-tested assistance	
	% of former wage (4)	*Duration (yrs.)* (5)	*Rate* (6)	*Duration (yrs.)* (7)
Belgium	40	1 + $^1/_4$ tenure	50% of min. wage	Indef.
Denmark	—	—	—	—
France	—	—	—	—
Germany	56	Indef.	—	—
Ireland	—	—	£28 p.w.	—
Italy	—	—	—	—
Netherlands	—	—	Dfl. 1045 p.m.	Indef.
Spain	{70 ($^1/_2$yr) {60 (1 yr)	1.5	80% of min. wage	1.5
UK	—	—	36% of lower quartile	Indef.
Australia	—	—	39% of $^3/_4$ ave. wage	Indef.
New Zealand	—	—	38% of $^3/_4$ ave. wage	Indef.
Canada	—	—	—	—
USA	—	—	—	—
Japan	—	—	—	—
Austria	—	—	27–54% of wage	Indef.
Finland	—	—	—	—
Norway	—	—	—	—
Sweden	—	—	30% of ave. wage[a]	0.5[a]
Switzerland	—	—	—	—

beneficiary is *ex post* equal to 35% of the *average* gross wage incl. bonus (*source*: Giorgio Brunello). *Norway*: provided for up to 1.5 years in any consecutive two calendar years. *Sweden*: UI a percentage of the former wage subject to a maximum.
[a] Sweden cols. (6) and (7) show KAS: this is flat rate equal to about 30% of average wage. *Not* means-tested; *not* available once UI expired.

over/

Sources: Mainly US Department of Health and Social Services, *Social Security Programs Throughout the World 1985 (Reserve Report no. 60)* and Eurostat, *Definition of Registered Unemployed*, 1987, Theme 3, Series E. See also OECD, *Employment Outlook*, Sept. 1988, Tables 4.3, 4.4. *Australia*: Gregory (1986); *New Zealand*: data from NZ High Commission in London; *Norway*: Strand (1986).

Table A2 *Unemployment benefits: coverage and generosity*

	(1) % receiving benefits (1986)	(2) Expenditure on benefits per unemployed person(as % of output per worker) (1987)	(3) Expenditure on benefits per recipient (as % of output per worker) (1986/7)	(4) Replacement ratio (%) (1985)
Belgium	85	18	22	60
Denmark	73	42	58	90
France	41	11	26	57
Germany	61	19	31	63
Ireland	67	15	23	50
Italy	21	4	20	2
Netherlands	—	26	—	70
Spain	35	9	—	80
UK	73	14	19	36
Australia	—	14	—	—
New Zealand	—	25	—	—
Canada	—	17	—	60
USA	34	9	26	50
Japan	40	15	36	60
Austria	—	23	—	60
Finland	—	19	—	75
Norway	—	17	—	65
Sweden	86/70	36	—	80
Switzerland	—	11	—	70

Notes: For the time–series of replacement ratios used in Ch. 11, see Emerson (1988: Appendix A). His numbers were scaled to give the same numbers as we use for 1985. Col. (3) is included for rough comparison with col. (4).

Sources: Col. (1): Eurostat, Series 3c *Labour Force Survey 1986*; *Japan*: From Giorgio Brunello; *Spain*: Bentolila and Blanchard (1990); *Sweden*: Burtless (1987: Table 70) reports 86%; Standing (1988) quotes AMS, Annual Report 1985/6, p. 16 as giving 70%; *UK*: *Employment Gazette*, Oct. 1988, p. 536, Table 1; *USA*: Burtless (1987: Table 8). (Burtless's figures for France, Germany, and UK are higher than here, especially for France, but for these countries his data represent all recipients ÷ all unemployed.) Col. (2): OECD, *Employment Outlook*, Sept. 1988, p. 86 gives expenditure on unemployment benefits as percentage of GDP in 1987. This is then divided by $u/(1 - u)$ from Table 2. This gives $(BU/Y)(N/U) = B/(Y/N)$. Col. (3): col. (2) ÷ col. (1). Col. (4): Table 5.

4. Wage-bargaining systems in OECD countries

Note: For coverage, 'high' = 75+ per cent, 'medium' = 25–75 per cent, 'low' = under 25 per cent.

Belgium

Coverage
High.

Bargaining system
Industry-level bargains applying by law to all workers in the sector.
Firm-level bargains supplement these.

Co-ordination
Low employer co-ordination and unions divided between Socialist, Christian, and liberal.
But central tripartite agreement, 1987–8.

Minimum wage
Set by national employer–union bargain.

Incomes policy
1982: government suspends indexation, and controls on permitted
   indexation last to 1986.

Denmark

See Scandinavia.

France

Coverage
High.

Bargaining system
Industry-level bargains applying by law to all workers in the sector.
Most pay above this determined at employer's discretion.

Co-ordination
Employers' confederation has some influence. Unions divided
   between four federations.

National minimum wage
Established by law. High.

Incomes policy
June 1982: wage and price freeze for three months followed by
   strict public-sector pay limits, generally followed by private sector at least as regards minimum rates in each grade (as a result
   of informal agreements with employers' organization).

Germany

Coverage
High.

Bargaining system
Industry-level bargains in each region, frequently extended by law
   to all workers in the sector.
Firm-level bargains supplement these but strikes at this level illegal.

Co-ordination
Informal talks within and between national employers' organization and trade union federation. This leads to a pattern settlement generally in the metal industry in one region, broadly
followed elsewhere. National industry-level union has to autho-
rize any strike. Council of Economic Experts and the five
research institutes help to create climate of opinion in favour of
wage moderation.

Minimum wage
No statutory level, except via collective bargaining.

Incomes policy
1966–77: 'Concerted action'. Tripartite guidance on wage limits.
Unions withdrew over 'co-determination'. Otherwise, see above.
Indexation illegal.

Ireland

Similar to UK. Wage accord 1979–81.

Italy

Coverage
High.

Bargaining system
Industry-level bargains applying by law to all workers in the sec-
tor.
Firm-level bargains supplement these.

Co-ordination
Some employer co-ordination, especially regionally. Strike insurance by employers. Union confederations have variable control
over their members (more in 1980s).

Minimum wage
See above.

Incomes policy
1976–8: 'historic compromise' establishing full indexation in
return for presumption of low settlements and reasonable strike
behaviour.

1983: agreement to alter calculation of *COL*.

1984: government proposal for reduction of permitted degree of indexation in the 'scala mobile'. Rejected by CGIL, union federation. Confirmed by plebiscite on 9 June 1985.

Netherlands

Coverage
High.

Bargaining system
Industry-level bargains applying by law to all workers in the sector.
Firm-level bargains supplement these, but at this stage strikes are outlawed.

Co-ordination
Three federations of employers and three of unions. More employer co-ordination than in Belgium. Since 1982 the Foundation of Labour, a joint employer-union organization, proposes a general framework for pay.

Minimum wage
Set by law.

Incomes policy
Tripartite incomes policy broke down in 1963. Frequent short wage freezes between 1971 and 1982.

Portugal

Coverage
High (despite very low unionization).

Bargaining system
Industry-level bargains applying by law to all workers in the sector. Unions have no strike funds.
No firm-level bargains but employers can pay at their discretion. Workers' committees in firms cannot negotiate.

Co-ordination
Three employers' confederations (for agriculture, industry, and trade). Two union confederations (Socialist and Communist). Permanent Social Concertation Council sets guidelines (since 1984).

Mediation
Government arbitration.

Minimum wage
Set by state. High relative to average earnings.

Incomes policy
1986–8: form of social contract to limit wage increases to expected inflation.

Spain

Coverage
High (despite low unionization).

Bargaining system
National tripartite framework agreement on permitted range of settlements (following 1978 Moncloa pact).
Industry-level bargains applying by law to all workers in the sector.
Firm-level bargains between firm and workers' committee (with mainly union members) supplement these. Some large firms do not participate in industry-level bargains.

Co-ordination
Two union movements (Socialist and Communist) in competition, with some co-ordination. Employers rather weak.

Minimum wage
Set by state.

Incomes policy
See above.

UK

Coverage
High (after including workers covered by statutory Wages Councils).

Bargaining system
Some industry-level bargains.
Majority of private-sector workers covered by firm-level bargains (sometimes building on industry-level bargains).

Co-ordination
Virtually none among employers. Ditto among unions.

Minimum wage
No national minimum (but Wages Council rates have force of law).

Incomes policy
1972–4: wage freeze (for six months); £1 + 4 per cent (for six months); 5 per cent plus extra if inflation exceeded 7 per cent (which it did). Statutory.
1975–9: £6 a week (1st year); 5 per cent (2nd year); 10 per cent (3rd year); 5 per cent (4th year). Supported by TUC in first two years.

Australia

Coverage
High.

Bargaining system
National Industrial Relations Commission (present title) sets general principles for pay increases. Industry-level bargains either follow these principles or have to be endorsed by the Commission. All such bargains relate to minimum rates. However, firm-level bargains can agree 'over-award' pay increases.

Co-ordination
Employers' federation generally weaker than union federation.

Incomes policy
1983 onwards: Prices and Incomes Accord between government and unions, ratified each year by the Commission (which has power to reject it). 'No extra claims' allowed at firm level—this being a key difference from the earlier regime. Firms remain free to make voluntary 'over-award' payments, and wage drift continues at 1.5–2 per cent p.a. Policy modified substantially from 1988 with a reversion to industry-level bargaining with 'extra claims' permitted—all being subject to Commission approval.

New Zealand

Coverage
Medium.

Bargaining system
Pre-1984: similar to Australia. 1984: compulsory arbitration abolished. Though Arbitration Commission continues to register most settlements, increasing proportion of settlements are made with no Commission involvement.

Co-ordination
As Australia.

Incomes policy
1971–84: wage and price controls of some kind for most of the period.

Canada

Coverage
Medium.

Bargaining system
Firm-level bargains. More public sector bargaining than in the USA.

Co-ordination
Nil.

Minimum wage
Set by state.

Incomes policy
1975–7: wage controls.

USA

Coverage
Low.

Bargaining system
Firm-level.

Co-ordination
Nil, though some pattern bargaining within industries.

Minimum wage
Set by government (low).

Incomes policy
1971: 90-day wage freeze, and controls lasting to 1974.
1978–9: Commission on Wage and Price Stability promotes pay
and price standards (essentially voluntary).

Japan

Coverage
Medium (high in large firms, low in small firms).

Bargaining system
Firm-level bargains, synchronized in Shunto (Spring offensive).

Co-ordination
Strong employer co-ordination, especially after the great inflation
of 1974. Weaker union co-ordination.

Incomes policy
Nil.

Austria

Coverage
High.

Bargaining system
Industry-level agreements, which depend on approval by the
union confederation.

Co-ordination
Strong guidance from tripartite Parity Commission and its bipar-
tite Subcommittee on Wages and Prices.

Incomes policy
None, as such.

Scandinavia

Coverage
High.

Bargaining system
National bargain between trade union federation and employers'
federation: one bargain in Denmark, three in Finland, Norway,
and Sweden. No Swedish bargain in 1990.
Industry-level and firm-level bargains supplement these, but
strikes are not allowed at firm level (because of peace agreements at higher level). National unions have to agree to local
claims. LO, manual union federation, controls strike fund
nationally in Sweden.

Co-ordination
Strong employers' and union federations, e.g. powerful co-ordination after 1982 Swedish devaluation.

Mediation
In Denmark and Norway this is compulsory, and there is sometimes binding arbitration.

Minimum wage
In Denmark industry-level minimum, set by state.

Incomes policies
Denmark: frequent legislative intervention setting ceiling on wage
growth; 1982: wage freeze; 1983: indexation suspended; 1985–7:
legal wage norms.
Finland: comprehensive tripartite incomes policy since 1968.
Wage indexation opposed.
Norway: frequent social contracts mainly in the 1970s. Firm-level
wage bargaining prohibited in 1978–9 and 1988.
Sweden: minimal direct intervention, though occasional guide-
lines.

Switzerland

Coverage
Medium.

Bargaining system
Mainly firm-level. Mainly subject to industry-wide five- to six-year

Annexes

peace agreements ruling out use of strikes. Multi-year settlements. Cost-of-living agreements negotiated at industry level.

Co-ordination
Strong employer co-ordination. Unions weak.

Arbitration
Important.

Sources: Blum (1981); Flanagan *et al.* (1983); Bruno and Sachs (1985); ILO (1987); OECD (1989); OECD *Country Reports*; Ashenfelter and Layard (1983); Calmfors and Driffill (1988); Calmfors (1990); Dore *et al.* (1989); Elvander (1989); Emerson (1988); miscellaneous country documents; numerous conversations, especially with Guillermo de la Dehesa, Ronald Dore, David Marsden, and, above all, David Soskice.

5. Optimal disinflation policy with hysteresis in wage-setting

We shall assume we wish to set the path of unemployment (u) to minimize

$$\int_0^\infty \frac{1}{2}(u^2 + \varphi\pi^2)e^{-rt}dt$$

$$\text{s.t. } \dot{\pi} = \theta_1(u^* - u) - \theta_{11}\dot{u},$$

where π is inflation and r the real discount rate. The differential equation for unemployment is

$$\ddot{u} - r\dot{u} - \delta(u - u^*) = 0,$$

where $\delta = [(\theta_1 + \theta_{11}r)\varphi\theta_1] / [(1 + \theta_{11}^2\varphi)]$. The stable solution to this equation is

$$u - u^* = Ae^{\alpha t},$$

where

$$\alpha = \frac{r}{2} - \sqrt{\left(\frac{r}{2}\right)^2 + \delta} < 0.$$

The speed of approach to u^* is $-\alpha$, which is increasing in δ and

$$\text{sign } \frac{\partial\delta}{\partial\theta_{11}} = \text{sign}[r(1 - \theta_{11}^2\varphi) - 2\theta_{11}\varphi\theta_1].$$

Thus, for small r, an increase in hysteresis (θ_{11}) decreases the speed of convergence to the NAIRU.

Suppose we start with inflation at an unacceptably high level. We need to go through a period of higher unemployment. Since convergence is monotonic, u goes straight to its maximal height. But the total fall in inflation is proportional to $\int(u_t - u^*)\mathrm{d}t$ (since $\int \dot{u}\mathrm{d}t = 0$). Hence slow convergence means that maximal u is low.

This analysis also applies to the case of accommodation to a temporary supply shock. Such a shock would generate a given amount of extra inflation if u were not raised. The optimal path of u to offset this extra inflation involves a smaller and more prolonged rise in u, the more hysteresis there is.

6. Unemployment and inflation series for each OECD country

Table A3 *Unemployment rates (%)*

	Australia	Austria	Belgium	Canada	Denmark	Finland	France
1955	0.8	2.9	4.1	4.0	4.4	0.5	2.1
1956	1.4	2.8	3.0	3.2	4.9	0.7	1.6
1957	2.0	2.6	2.4	4.3	4.6	1.8	1.2
1958	2.6	2.8	3.5	6.5	4.2	3.1	1.4
1959	2.5	2.5	4.2	5.5	3.0	2.2	1.9
1960	1.9	1.9	3.4	6.3	2.4	1.5	1.8
1961	3.5	1.5	2.6	6.5	2.1	1.2	1.5
1962	2.8	1.5	2.2	5.4	2.1	1.3	1.4
1963	2.7	1.7	1.8	5.0	2.3	1.5	1.3
1964	1.7	1.6	1.6	4.3	1.9	1.5	1.4
1965	1.5	1.6	1.8	3.6	1.7	1.4	1.5
1966	1.7	1.5	2.0	3.3	1.9	1.5	1.8
1967	1.9	1.6	2.6	3.8	1.7	2.9	1.9
1968	1.8	1.6	3.1	4.4	1.7	3.8	2.6
1969	1.8	1.6	2.3	4.4	1.7	2.8	2.3
1970	1.6	1.1	2.1	5.6	1.3	1.9	2.5
1971	1.9	1.0	2.1	6.1	1.6	2.2	2.7
1972	2.6	1.0	2.7	6.2	1.6	2.5	2.8
1973	2.3	1.0	2.7	5.5	1.0	2.3	2.7
1974	2.6	1.2	3.0	5.3	2.3	1.7	2.8
1975	4.8	1.5	5.0	6.9	5.3	2.2	4.0
1976	4.7	1.6	6.4	7.1	5.3	3.8	4.4
1977	5.6	1.4	7.4	8.0	6.4	5.8	4.9
1978	6.2	1.8	7.9	8.3	7.3	7.2	5.2
1979	6.2	1.8	8.2	7.4	6.2	5.9	5.9
1980	6.0	1.6	8.8	7.5	7.0	4.6	6.3
1981	5.7	2.2	10.8	7.5	9.2	4.8	7.3
1982	7.1	3.1	12.6	10.9	9.8	5.3	8.1
1983	9.9	3.7	12.1	11.8	10.4	5.4	8.3
1984	8.9	3.8	12.1	11.2	10.1	5.2	9.7
1985	8.2	3.6	11.3	10.4	9.0	5.0	10.2
1986	8.0	3.1	11.2	9.5	7.8	5.3	10.4
1987	8.0	3.8	11.0	8.8	7.8	5.0	10.5
1988	7.2	3.6	9.7	7.7	8.6	4.5	10.0
1989	6.1	3.2	8.0	7.5	9.5	3.4	9.4
1990	6.9	3.2	7.2	8.1	9.6	3.4	8.9
1991	9.5	3.5	7.1	10.2	10.5	7.5	9.4
1992	10.7	3.6	7.8	11.2	11.3	13.0	10.2

	Germany	Ireland	Italy	Japan	Netherlands	Norway	New Zealand
1955	4.3	4.6	7.0	2.7	1.4	1.7	0.0
1956	3.5	5.3	8.6	2.5	1.0	2.0	0.0
1957	2.9	6.7	7.2	2.0	1.3	2.1	0.1
1958	3.0	6.4	6.3	2.1	2.4	3.4	0.1
1959	2.0	6.1	5.7	2.2	1.9	3.3	0.1
1960	1.1	5.6	4.4	1.7	1.2	2.4	0.1
1961	0.6	5.1	3.8	1.5	0.9	1.8	0.0
1962	0.6	4.9	3.3	1.3	0.8	2.1	0.1
1963	0.4	5.0	2.9	1.3	0.8	2.5	0.1
1964	0.4	4.7	3.2	1.2	0.7	2.2	0.1
1965	0.3	4.6	4.2	1.3	0.8	1.8	0.1
1966	0.2	4.7	4.4	1.4	1.1	1.6	0.0
1967	1.3	5.0	4.0	1.3	2.1	1.5	0.4
1968	1.5	5.3	4.2	1.2	2.0	2.1	0.7
1969	0.9	5.0	4.2	1.1	1.3	2.0	0.3
1970	0.8	5.8	3.8	1.1	1.3	1.6	0.1
1971	0.9	5.5	3.9	1.2	1.7	1.5	0.3
1972	0.8	6.2	4.5	1.4	2.9	1.7	0.5
1973	0.8	5.7	4.4	1.3	2.9	1.5	0.2
1974	1.6	5.3	3.7	1.4	3.6	1.5	0.1
1975	3.6	7.3	4.0	1.9	5.2	2.3	0.3
1976	3.7	9.0	4.6	2.0	5.5	1.8	0.4
1977	3.6	8.8	4.9	2.0	5.3	1.5	0.6
1978	3.5	8.2	4.9	2.2	5.3	1.8	1.7
1979	3.2	7.1	5.2	2.1	5.4	2.0	1.9
1980	3.0	7.3	5.2	2.0	6.0	1.6	2.7
1981	4.4	9.9	5.8	2.2	8.5	2.0	3.5
1982	6.1	11.4	6.4	2.4	11.4	2.6	3.7
1983	8.0	14.0	7.0	2.6	12.0	3.4	5.4
1984	7.1	15.5	7.0	2.7	11.8	3.1	4.6
1985	7.2	17.4	7.1	2.6	10.6	2.6	3.6
1986	6.4	17.4	7.5	2.8	9.9	2.0	4.0
1987	6.2	17.5	7.9	2.8	9.6	2.1	4.1
1988	6.2	16.7	7.9	2.5	9.2	3.2	5.6
1989	5.6	14.7	7.7	2.3	8.3	4.9	7.1
1990	4.9	13.4	7.0	2.1	7.5	5.2	7.8
1991	4.4	14.9	6.9	2.1	7.0	5.5	10.3
1992	4.8	16.1	6.9	2.2	6.7	5.9	10.3

	Spain	Sweden	Switzerland	UK (1)	UK (2)	USA
1955	2.3	1.7	1.0	1.4	—	4.3
1956	2.1	1.3	1.1	1.5	—	4.0
1957	1.8	1.6	0.7	1.9	—	4.1
1958	1.7	2.2	1.2	2.8	—	6.6
1959	1.9	1.8	0.8	2.9	—	5.3
1960	2.5	1.3	0.4	2.2	—	5.3
1961	2.5	1.1	0.2	1.9	—	6.4
1962	2.0	1.1	0.2	2.9	—	5.3
1963	2.2	1.4	0.2	3.5	—	5.5
1964	2.6	1.2	0.0	2.6	—	5.0
1965	2.5	1.0	0.0	2.2	—	4.4
1966	2.1	1.3	0.0	2.2	—	3.6
1967	2.5	1.7	0.0	3.1	—	3.7
1968	3.0	1.8	0.0	3.1	—	3.5
1969	2.6	1.5	0.0	2.9	—	3.4
1970	2.4	1.2	0.0	3.0	—	4.8
1971	3.1	2.1	0.0	3.8	2.6	5.8
1972	3.1	2.2	0.0	4.2	2.9	5.5
1973	2.5	2.0	0.0	3.1	2.0	4.8
1974	2.6	1.6	0.0	3.1	2.0	5.5
1975	3.7	1.3	0.9	4.5	3.1	8.3
1976	4.7	1.3	1.8	5.7	4.2	7.6
1977	5.2	1.5	1.2	6.1	4.4	6.9
1978	6.9	1.8	0.9	5.9	4.3	6.0
1979	8.5	1.7	0.9	5.0	4.0	5.8
1980	11.2	1.6	0.6	6.4	5.1	7.0
1981	13.9	2.1	0.6	9.8	8.1	7.5
1982	15.8	2.6	1.2	11.3	9.5	9.5
1983	17.2	2.9	2.4	12.4	10.5	9.5
1984	20.0	2.6	3.0	11.7	10.7	7.4
1985	21.4	2.4	2.4	11.2	10.9	7.1
1986	21.0	2.2	2.1	11.2	11.1	6.9
1987	20.1	1.9	1.8	10.3	10.0	6.1
1988	19.1	1.6	2.1	8.6	8.1	5.4
1989	16.9	1.4	1.8	7.2	6.3	5.2
1990	15.9	1.5	1.8	6.8	5.8	5.4
1991	16.0	2.7	3.9	8.7	8.1	6.6
1992	18.1	4.8	9.0	10.0	9.8	7.3

Notes: Standardized rates except for Denmark, Ireland, New Zealand, Austria, and Sweden, for which unstandardized rates are used. The standardized unemployment rates are described in 'Who are the unemployed? Measurement issues and their policy implications', OECD, *Employment Outlook*, Sept. 1987, pp. 125–41; and in C. Sorrentino, 'The Uses of the Community Labour Force Surveys for International Unemployment Comparisons', Eurostat Document no. 7 for Seminar on 12–14 Oct. 1987. Except for Italy, these numbers are very similar to the 'unemployment rate on US concepts', calculated by the US Bureau of Labor Statistics: see 'Comparative Labor Force Statistics for 10 Countries 1959–88' (mimeo).

For Italy we use the BLS numbers 'on US concepts', which exclude the considerable number of Italian people who, though registered as unemployed, have performed no active job search in the previous 4 weeks. For 1985 and earlier we multiply the BLS numbers by 7.5/6.3 to allow for the break. (See p. 2 of the document.)

For Switzerland we use registered unemployment × 3, this being the factor for 1980 shown in the 1980 Census.

For the UK we give two series: (1) a series based on OECD data and (2) the UK Department of Employment's consistent series.

Further details on request.

Sources: *EC*: OECD, *Economic Outlook*, Dec. 1990, Tables R18, R19 (updated using Tables 40, 41), except for Italy, Switzerland, and the UK (see Notes). Recent data is taken from OECD *Quarterly Labour Force Statistics*, No. 1, 1993 and *Main Economic Indicators*, April 1993.

Table A4 *Inflation rate (% p.a.)*

	Australia	Austria	Belgium	Canada	Denmark	Finland	France
1955	3.2	3.1	1.5	0.6	4.8	2.9	2.1
1956	6.9	4.1	3.3	3.6	4.8	9.0	4.9
1957	0.1	4.3	4.1	2.2	1.8	7.4	6.0
1958	0.0	0.5	1.0	1.4	1.6	7.8	11.8
1959	4.5	3.6	0.4	2.0	3.8	1.3	6.3
1960	3.2	3.1	0.8	1.2	1.8	2.2	3.5
1961	1.7	5.5	1.1	0.4	4.8	5.6	3.7
1962	1.1	3.9	1.8	1.2	6.5	3.8	4.1
1963	2.8	3.4	3.2	2.0	5.5	5.1	6.8
1964	3.2	3.3	4.5	2.7	4.6	7.6	4.3
1965	2.6	5.6	5.0	3.4	7.7	5.1	2.6
1966	3.6	3.1	4.4	4.8	6.7	4.3	3.0
1967	2.5	3.5	3.0	4.2	6.3	7.6	2.9
1968	3.8	2.6	2.6	3.7	7.2	12.0	4.2
1969	4.6	2.8	4.0	4.6	6.8	4.4	6.8
1970	4.0	4.9	4.7	4.4	8.3	3.7	5.5
1971	6.4	6.1	5.5	3.3	7.7	7.6	6.4
1972	7.6	7.5	6.2	5.8	9.2	8.3	7.5
1973	11.9	8.0	7.3	8.7	10.9	14.2	8.4
1974	18.2	9.5	12.7	14.5	12.9	22.6	12.3
1975	16.0	6.4	12.0	9.9	12.4	14.3	13.2
1976	13.3	5.7	7.6	8.6	9.3	12.5	10.9
1977	9.2	5.1	7.6	6.4	9.3	10.3	8.9
1978	7.6	5.7	4.3	5.9	10.0	8.1	10.1
1979	10.0	4.0	4.6	10.3	7.6	8.4	10.1
1980	11.3	5.2	3.7	10.5	8.3	9.6	11.4
1981	9.5	6.5	4.8	10.8	10.0	11.8	11.5
1982	11.1	6.2	7.1	8.7	10.6	8.5	11.7
1983	8.1	3.9	5.5	5.0	7.6	8.6	9.77
1984	6.3	4.9	5.3	3.2	5.6	8.9	7.4
1985	6.4	3.1	6.0	2.6	4.4	5.2	5.8
1986	7.3	4.3	3.8	2.4	4.6	4.6	5.2
1987	7.4	2.4	2.3	4.7	4.7	5.3	3.0
1988	8.3	1.6	1.8	4.7	3.4	6.9	2.9
1989	7.3	2.9	4.7	4.8	4.2	6.9	3.1
1990	4.6	2.9	2.7	3.3	2.5	5.4	3.0
1991	1.8	3.4	2.7	2.7	2.5	2.3	3.0
1992	1.1	4.3	3.0	1.0	1.9	1.1	2.3

	Germany	Ireland	Italy	Japan	Netherlands	Norway	New Zealand
1955	2.1	2.3	3.3	1.6	4.5	4.5	1.5
1956	3.0	2.8	4.0	5.0	3.9	7.5	3.1
1957	3.1	3.7	2.0	6.3	5.7	3.5	1.8
1958	3.4	5.8	2.4	−1.7	1.8	0.8	1.4
1959	1.4	2.6	−0.3	2.9	2.0	0.8	2.9
1960	2.5	0.1	2.1	6.0	2.7	1.0	2.1
1961	5.1	3.1	2.4	7.9	5.3	2.7	−3.9
1962	3.8	5.1	5.9	4.0	0.4	4.7	6.5
1963	3.1	1.9	8.9	5.7	5.0	3.0	1.5
1964	3.0	10.4	6.1	5.4	8.5	4.8	4.5
1965	3.9	4.3	3.9	5.1	6.1	5.1	2.2
1966	3.3	4.1	2.8	4.8	6.1	4.0	−2.8
1967	1.6	3.2	2.7	5.5	4.2	3.0	8.7
1968	2.2	4.6	1.8	4.9	4.1	4.1	6.0
1969	4.2	9.5	3.5	4.4	6.4	4.3	0.6
1970	7.6	9.3	7.5	6.5	6.3	12.8	13.8
1971	7.8	10.4	7.0	5.6	8.1	6.7	11.5
1972	5.3	13.8	5.8	5.8	9.3	5.0	9.9
1973	6.2	15.1	13.7	13.2	9.0	9.3	9.4
1974	7.0	5.9	19.9	20.0	9.2	10.1	2.9
1975	5.9	20.3	16.1	7.4	10.2	10.2	13.9
1976	3.6	21.2	18.6	7.8	8.9	7.4	18.9
1977	3.7	13.1	18.6	6.4	6.7	8.4	16.8
1978	4.3	10.6	13.9	5.0	5.5	6.4	12.6
1979	3.8	13.5	15.4	2.7	3.9	6.6	14.3
1980	4.9	14.8	20.1	4.5	5.6	14.5	15.9
1981	4.1	17.4	18.9	3.8	5.6	14.0	15.7
1982	4.4	15.1	17.2	1.6	6.1	10.2	10.7
1983	3.6	10.8	15.1	1.5	1.9	6.2	7.5
1984	2.0	6.4	11.7	2.3	1.9	6.4	8.0
1985	2.3	5.2	8.8	1.5	1.8	5.0	13.6
1986	3.3	6.5	7.9	1.8	0.5	−1.4	18.2
1987	1.9	2.4	5.9	0.0	−0.6	7.1	11.7
1988	1.5	2.8	6.7	0.4	1.2	4.4	9.1
1989	2.6	4.7	6.2	1.9	1.2	5.9	7.2
1990	3.5	−1.5	7.6	2.2	2.5	4.6	3.3
1991	4.1	1.2	7.4	2.1	3.0	2.0	1.9
1992	4.6	6.1	4.7	1.8	1.9	−1.0	2.4

	Spain	*Sweden*	*Switzerland*	*UK*	*USA*
1955	6.1	4.4	1.0	4.2	1.8
1956	7.2	5.2	0.8	6.4	3.1
1957	12.4	4.3	2.3	3.7	3.3
1958	10.0	3.2	5.1	4.0	1.8
1959	5.8	1.1	−0.1	1.3	2.1
1960	0.5	4.9	2.8	1.0	1.5
1961	2.6	2.9	4.1	2.3	0.7
1962	5.1	4.4	5.8	3.8	2.5
1963	8.5	2.7	4.9	2.2	1.1
1964	6.7	4.2	5.2	3.6	1.7
1965	9.5	6.0	3.9	5.5	2.7
1966	7.7	6.6	4.8	4.6	3.7
1967	8.0	5.3	4.6	2.5	2.9
1968	5.0	2.1	3.0	4.3	5.0
1969	4.7	3.7	2.7	5.3	5.0
1970	6.8	5.1	4.6	7.8	5.4
1971	7.8	7.1	9.1	9.3	5.4
1972	8.5	7.0	9.9	8.0	4.8
1973	12.1	6.9	8.1	7.4	6.3
1974	16.1	9.5	6.9	14.6	8.7
1975	16.7	14.6	7.2	27.3	9.7
1976	16.7	11.7	2.8	15.3	6.1
1977	23.1	10.7	0.3	13.8	6.9
1978	20.7	9.5	3.5	11.5	7.9
1979	16.9	7.9	2.0	14.3	8.8
1980	14.3	11.7	3.0	19.5	9.4
1981	12.0	9.6	6.9	11.3	10.0
1982	13.8	8.2	7.1	7.6	6.2
1983	11.6	10.1	2.9	5.2	4.1
1984	11.0	7.6	2.8	4.7	4.4
1985	8.6	6.6	3.1	5.7	3.6
1986	11.1	6.9	3.8	3.5	2.7
1987	5.8	4.7	2.6	4.9	3.2
1988	5.7	6.5	2.4	6.6	3.9
1989	7.0	8.1	4.2	7.1	4.5
1990	7.4	8.9	5.6	6.2	4.4
1991	6.8	7.5	6.1	6.5	4.0
1992	5.3	1.6	2.4	4.5	2.6

Sources: GNP/GDP deflator: OECD, *Economic Outlook*, various issues, and Centre for Economic Performance, 'OECD dataset'.

Discussion Questions

1. Consider the following statements:
   (i) Changes in inflation depend on the level of unemployment.
   (ii) Inflation depends on financial factors and in particular on the growth of national spending in nominal terms.
   What truth, if any, is there in either view?
2. 'Unemployment is high because real wages are too high.' Discuss.
3. Consider a wholly unionized economy. It is argued that, 'if each wage is set by a voluntary contract between a firm and its union, the outcome must be efficient.' Is this logically correct? Might it be preferable to have a single national bargain between the employers' federation and the union federation? If so, why?
4. 'If there are no unions, there is no reason why there should be involuntary unemployment.' Discuss.
5. 'Productivity growth can have no effect on unemployment.' Discuss. Is the same true of changes in taxes and the terms of trade?
6. 'If unemployed people look harder for work, this cannot affect the number of jobs.' Discuss.
7. 'There are always some jobs available, so unemployment cannot be due to job rationing.' Discuss.
8. How would you explain the different unemployment rates in different countries?
9. What policies, if any, would reduce unemployment? Even if they would, would they be desirable in some overall sense?

More jobs wanted than exist.

List of Symbols

—   Not available

Variables

P	price of value added
P_c	price of consumption
P_m	price of imports
W	wage rate including employers' taxes
Z	'push factors' affecting wages
L	labour force
N	employment
K	capital
Y	output
t_1	tax rate on wages, paid by employer
t_2	tax rate on wages, paid by worker
t_3	indirect tax rate
E	effort (also written $e(\cdot)$)
Π	profit
A	alternative expected income (if disemployed)
U	unemployed (number)
u	unemployment rate, U/L
u^*	NAIRU
u_s^*	short-run NAIRU
c	search effectiveness
V	vacancies (number)
v	vacancy rate, V/N
H	hirings, number per period
s	separation rate, fraction of employed workers who become unemployed per period.

Where other uses occur, this is made clear. For the first nine variables, the logarithm is denoted by the use of lower-case letters.

Key recurring parameters come from the following relations

Prices: $\quad\quad p - w = \beta_0 - \beta_1 u - \beta_{11}\Delta u - \beta_2\Delta^2 p.$
Wages: $\quad\quad w - p = \gamma_0 - \gamma_1 u - \gamma_{11}\Delta u - \gamma_2\Delta^2 p = z.$
Production: $\quad y = \alpha n + (1 - \alpha)k.$
Demand: $\quad\quad y_i = -\eta p_i.$

Estimation

Throughout the book, we omit equation diagnostics. In general, all regression equations have serially uncorrelated errors and stable parameters. Figures in brackets are normally *t*-statistics.

Statistics

All figures are individually rounded and therefore do not necessarily add up.

Germany

Throughout the book 'Germany' refers to 'West Germany'.

List of Figures

References

ABOWD, J. M. and CARD, D. M. (1989), 'On the covariance structure of earnings and hours changes', *Econometrica*, 57: 411–46.

ALTONJI, J. G. (1982), 'The Intertemporal Substitution Model of Labour Market Fluctuations: An Empirical Analysis', *Review of Economic Studies*, 49(5): 783–824.

ALTONJI, J. G. (1986), 'Intertemporal Substitution in Labor Supply: Evidence from Micro Data', *Journal of Political Economy*, 94(3): S176–S215.

ALTONJI, J. G. and ASHENFELTER, O. (1980), 'Wage Movements and the Labour Market Equilibrium Hypothesis', *Econometrica*, 47: 217–45.

ASHENFELTER, O. (1984), 'Macroeconomic Analyses and Microeconomic Analyses of Labor Supply', in K. Brunner and A. Meltzer (eds.), *Essays on Macroeconomic Implications of Financial and Labor Markets and Political Processes, Carnegie–Rochester Conference Series on Public Policy*, xxii, pp. 117–55.

ASHENFELTER, O. C. and LAYARD, R. (1983), 'Incomes Policy and Wage Differences', *Economica*, 50: 127–43.

ATKINSON, A. B. and MICKLEWRIGHT, J. (1991), 'Unemployment Compensation and Labor Market Transitions: A Critical Review', *Journal of Economic Literature*, 29(4).

BARRO, R. J. and GROSSMAN, H. I. (1971), 'A General Disequilibrium Model of Income and Employment', *American Economic Review*, 61(1): 82–93.

BENTOLILA, S. and BLANCHARD, O. J. (1990), 'Spanish Unemployment', *Economic Policy*, no. 10: 234–81.

BINMORE, K., RUBINSTEIN, A. and WOLINSKY, A. (1986), 'The Nash Bargaining Solution in Economic Modeling', *RAND Journal of Economics*, 17(2): 176–88.

BJÖRKLUND, A. (1990), 'Evaluations of Swedish Labor Market Policy', Swedish Institute for Social Research, University of Stockholm, *Finnish Economic Papers*, 3(1).

BLINDER, A. S. (1979), *Economic Policy and the Great Stagflation*, New York: Academic Press.

BLUM, A. A. (ed.) (1981), *International Handbook of Industrial Relations*, London: Aldwych Press.

148 *References*

BRUNO, M. and SACHS, J. D. (1985), *Economics of Worldwide Stagflation*, Oxford: Basil Blackwell.

BURNSIDE, C., EICHENBAUM, M. and REBELO, S. (1993), 'Labor Hoarding and the Business Cycle', *Journal of Political Economy*, 101(2): 245–73.

BURTLESS, G. (1987), 'Jobless Pay and High European Unemployment' in R. Z. Lawrence and C. Schultze (eds.), *Barriers to European Growth: A Transatlantic View*, Washington DC: Brookings Institution.

CALMFORS, L. (ed.) (1990), *Wage Formation and Macroeconomic Policy in the Nordic Countries*, Oxford: Oxford University Press.

CALMFORS, L. and DRIFFILL, J. (1988), 'Centralisation of Wage Bargaining and Macroeconomic Performance', *Economic Policy*, no. 6: 13–61.

CRISTINI, A. (1989), *OECD Activity and Commodity Prices*, Lincoln College, Oxford, unpublished D.Phil. thesis.

DORE, R., BOUNINE-CABALÉ, J., and TAPIOLA, K. (1989), *Japan at Work: Markets, Management and Flexibility*, Paris: OECD.

ELVANDER, N. (1989), 'Incomes Policies in the Nordic Countries', mimeo.

EMERSON, M. (1988), *What Model for Europe?*, Cambridge, Mass.: MIT Press.

FEINSTEIN, C. H. (1972), *National Income, Expenditure and Output of the United Kingdom 1855–1965*, Cambridge: Cambridge University Press.

FLANAGAN, R. J., SOSKICE, D. W., and ULMAN, L. (1983), *Unionism, Economic Stabilization, and Incomes Policy: European Experience*, Washington DC: Brookings Institution.

FRIEDMAN, M. (1968), 'The Role of Monetary Policy', *American Economic Review*, 58: 1–17.

GREGORY, R. G. (1986), 'Wages Policy and Unemployment in Australia', *Economica*, 53: S53–S74. Reprinted in Bean *et al.* (1987).

HAM, J. C. (1986), 'Testing whether Unemployment Represents Intertemporal Labour Supply Behaviour', *Review of Economic Studies*, 53(4): 559–78.

HANSEN, G. D. (1985), 'Indivisible Labor and the Business Cycle', *Journal of Monetary Economics*, 16(3): 309–27.

ILO (International Labour Office) (1987), *World Labor Report, iii, Incomes from Work: Between Equity and Efficiency*, Geneva: ILO.

LAYARD, R. and PHILPOTT, J. (1991), *Stopping Unemployment*, London: Employment Institute.

LAYARD, R., NICKELL, S., and JACKMAN, R. (1991), *Unemployment: Macroeconomic Performance and the Labour Market*, Oxford: Oxford University Press.

LUCAS, R. E. Jr. (1972), 'Expectations and the Neutrality of Money', *Journal of Economic Theory*, 4(2): 103–24.

LUCAS, R. E. Jr. and RAPPING, L. A. (1969), 'Real Wages, Employment, and Inflation', *Journal of Political Economy*, 77(5): 721–54.

MALINVAUD, E. (1977), *The Theory of Unemployment Reconsidered*, Oxford: Basil Blackwell.

MANKIW, N. G., ROTEMBERG, J. J., and SUMMERS, L. H. (1985), 'Intertemporal Substitution in Macroeconomics', *Quarterly Journal of Economics*, 100(1): 225–51.

MILLWARD, N. and STEVENS, M. (1986), *British Workplace Industrial Relations 1980–1984: The DE/ESRC/PSI/ACAS Surveys*, Aldershot: Gower Press.

NASH, J. F. Jr. (1950), 'The Bargaining Problem', *Econometrica*, 18(2): 155–62.

NASH, J. F. Jr. (1953), 'The-Person Cooperative Games', *Econometrica*, 21(1): 128–40.

OECD (1989), *Economies in Transition*, Paris: OECD.

ROGERSON, R. (1988), 'Indivisible Labor, Lotteries and Equilibrium', *Journal of Monetary Economics*, 21(1): 3–16.

STANDING, G. (1988), *Unemployment and Labour Market Flexibility*, Geneva: International Labour Office.

STRAND, J. (1986), 'En Vurdering av Ordningen for Okonomisk Stotte til Permitterte Arbeidstakere i Norge', Norwegian Department of Labour, mimeo.

WADHWANI, S. (1985), 'Wage Inflation in the United Kingdom', *Economica*, 52: 195–207.

WEITZMAN, M. L. (1984), *The Share Economy*, Cambridge, Mass.: Harvard University Press.

INDEX